WAVES, TIDES, AND CURRENTS *hold a fascination for almost everyone. Elizabeth Clemons' clear, informative text gives concise background material which will serve as an excellent introduction for anyone who wants to understand the mysteries and wonders of the seas.*

Aided by photographs and Guy Fleming's maps and diagrams, Elizabeth Clemons describes the normal changes in the moods of the sea, the causes of the terrifying hurricanes and tidal waves, and the recent developments in using the resources of the oceans.

Waves, Tides, and Currents

Waves, Tides,

ILLUSTRATED WITH MAPS,

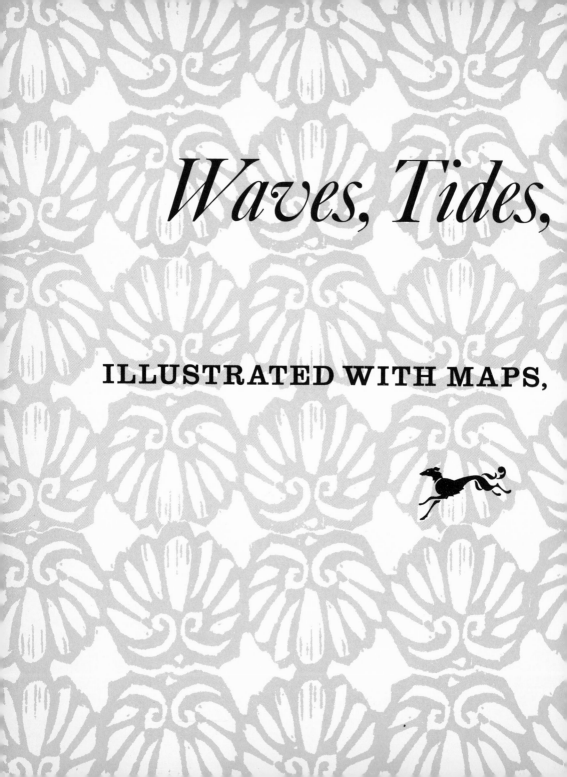

and Currents

by Elizabeth Clemons

DIAGRAMS & PHOTOGRAPHS

Alfred·A·Knopf / **NEW YORK**

THIS IS A BORZOI BOOK

PUBLISHED BY ALFRED A. KNOPF, INC.

Trade Ed.: ISBN: 0-394-81824-5   Lib. Ed.: ISBN: 0-394-91824-X

Manufactured in the United States of America

This title was originally catalogued by the Library of Congress as follows:

Clemons, Elizabeth.
  Waves, tides, and currents. New York, Knopf [1967]
  112 p. illus., map. 22 cm.
  Bibliography: p. 107–108.

1. Ocean waves—Juvenile literature. 2. Tides—Juvenile literature.
3. Ocean currents—Juvenile literature.   I. Title.

GC303.C55   j 551.4   66–13786

Library of Congress   [3]

To my husband
ARTHUR GRANVILLE ROBINSON
Vice Admiral United States Navy (Retired)
without whose encouragement and assistance this book
never would have been written.

Foreword

IF YOU HAVE STOOD BY THE OCEAN and wondered about the never-ending waves that come rolling in, this book is for you. It was written for boys and girls, and adults, too, who want to know more about waves, tides, and currents.

Although the book is not fiction, certain parts read almost as if it were a fairy tale instead of the truth. Much of science often seems too amazing to be true!

A few years ago some of the facts in this book were not well known. Others were questioned. Even now we do not have all the answers, although in recent years our knowledge of the sea has greatly enlarged.

Perhaps some of you will someday add to our knowledge about the world of water that surrounds us.

E.C.C., *Carmel Valley, California*

Acknowledgments

No book of this kind can be written without the help of many persons. Although full responsibility for any error in fact or in interpretation must be mine, the generous help of so many persons is greatly appreciated.

I am indeed grateful to the officers and professors of the United States Naval Post Graduate School in Monterey, California, who have assisted in making this an accurate account of waves, tides, and currents: to Rear Admiral Edward J. O'Donnell and to Professor Jacob Wickham, who read the finished manuscript and made vital and helpful suggestions; to Dr. Warren C. Thompson, who reviewed the first outline and made numerous changes in the original structure of the book; to Professor Carl Menneken, Dean of Research Administration, who assisted in many ways with firsthand information, gained through his work at Woods

Hole and on trips aboard the *Atlantis II;* and to Commander Richard Haupt, who provided information and illustrative materials used in checking sea water today.

Because reference materials and assistance in evaluating them are essential in writing books, my appreciation goes to the Harrison Memorial Library staff of Carmel, California, and to the Naval Post Graduate Library members, who helped to find necessary references; and to my many friends, who gave information and help, especially to Eugene M. Rice of Washington, D. C. for sources and information on storm damage; to Lawrence E. Thomas, of Morro Bay, California, who checked the tidal references for the Estero Bay District; and to Dorothy Beard, who typed the final manuscript of the book.

To my nephew, Bruce Cameron Marshall, who assisted in accurate interpretation of Northern California tides, and to my husband, Vice Admiral Arthur G. Robinson, USN Retired, who read the manuscript and helped in rewriting certain sections, I wish to express my heartfelt thanks.

And after a book is written, the editors with whom an author works give tremendous assistance. My sincere appreciation goes to Virginie Fowler, my friend and editor, and to the members of her staff, Jane Yolen Stemple and Shanna McNeill, for their careful reading and many suggestions for a better manuscript.

ELIZABETH CLEMONS, *Carmel Valley, California*

Contents

I · *Tides at the Seashore*   *3*

II · *The Earth, the Moon, and the Sun*   *9*

III · *What Causes the Tides?*   *1 9*

IV · *Ranges and Differences in Tides*   *2 7*

V · *Waves on the Shore and at Sea*   *4 3*

VI · *Currents of the Ocean: The Rivers of the Sea*   *6 5*

VII · *Waves, Tides, and Currents of Tomorrow*   *8 5*

Glossary   *1 0 5*

Bibliography   *1 0 7*

Photo Credits   *1 0 8*

Index   *1 0 9*

Waves, Tides, and Currents

CHAPTER I

Tides at the Seashore

ALONG THE EDGES OF THE OCEAN, water crashes toward shore! If you live near the sea, you know that the water sometimes comes high up on the shore. At other times the ocean is quite far out on the beach. These daily forward and backward motions of the sea are the high tides and the low tides.

Day after day, year after year, tides come in and go out, constantly following the same pattern over and over. Windstorms make the seas temporarily rise higher on beaches and against cliffs. However, the constant rise and fall of the tides follows a steady routine. The daily rise and fall of the sea surface is a continuing and never-ending process.

In most places, for approximately six continuous hours of any given day, the water of the sea moves in toward land. As it comes in closer to shore, the water rises higher and higher on the beach. This incoming water is called the *flood tide*. When the water stops rising, it is *high tide*.

Then the water moves away from shore. It "turns," or starts to go out. This is called the *ebb tide*. At the end of the next six hours, the water stops going out. Once the water reaches its lowest point, it is *low tide*.

As soon as low tide is reached, the water starts to move toward shore again. The change from low tide to high tide is not sudden. It is steady and constant. The same vertical rise and fall and the horizontal flow-and-ebb pattern of the moving sea water is repeated during the next twelve hours and twenty-five minutes of the day or night.

There are two high tides and two low tides each twenty-four hours in most places along the ocean. In some places the difference between high tide and low tide is less than one foot. In other places it may be as great as fifty feet. The difference between the water levels at high tide and low tide is called the *range* of the tides.

The time between the two high tides is about twelve hours and twenty-five minutes. If we have high tide at 7 a.m. on Monday morning, the next high tide will be that night at 7:25 p.m. Then the next high tide will be Tuesday morning at 7:50 a.m. Every twenty-four hours the tide is about fifty minutes later. Many people think of it as about an hour later each day.

The tide comes in and goes out in a continuous, unending pattern. There are two high and two low tides each day.
When the tide goes out, masses of seaweed that lie between the rocks along the shore are uncovered [below].

Tides are very noticeable in an area of shallow water. At low tide you may see just a wide strip of mud flats. When the tide comes in, the entire mud flat will no longer be land but will actually be covered with water. This often happens where there is low land or a flat area near the sea. The range of the tide in shallow water spreads out over a wide flat area.

If there is a sloping beach along the shore, the water runs up on the beach. It comes in higher and higher and then recedes or goes back toward the sea. You can see the mark of the highest part of the water on the damp sand. Broken shells, seaweeds, and small pieces of driftwood mark the place of the high tide.

Where the bits of material are left by the outgoing tide, you will be able to see that the water has run up higher on some places on the beach. This is because the water had more force or energy when it came in at that particular place.

Notice these various places as you go along the beach and try to figure out why the range of the tide is not in a straight line on the beach where you are walking.

Where the coastal area is steep, with cliffs near the beach, the range of tide is seen to be quite different. The tides along these areas make a definite water line on the straight sides of the cliffs. These marks show the actual rise in tide-water. The sea comes up on the cliff, covering the rocks along the shore. The sea rises in height against the cliff. It

is not spread out horizontally over land or mud flats. It does not run up on a sloping beach. The range of a tide against a cliff is almost the same as mercury rising in a thermometer. You can see the height of the tidewater on the cliff or on steep rocky shores, just as you can see the line rising with warmer temperatures on a thermometer.

Where the shore is rocky and extends into the sea, the tide comes in and fills small tide pools and covers rocky boulders.

When the tide goes out, the tide pools, rocks, and masses of seaweed that lie between the rocks along the shore are uncovered. The range of the tide can often be judged by how much of the rock is covered, or how completely the tide pools are filled at high tide.

Even if the rocky shore does not extend into the sea, waves with wind behind them are strong enough to crash over high rocky areas. Use caution in walking near the edges of such rocky places. Waves come in with great force.

CHAPTER II

The Earth, the Moon, and the Sun

To understand the tides, you must know something about the earth, the moon, and the sun.

The earth moves around the sun once a year. The path of the earth about the sun is called its *orbit*. The journey takes a few minutes less than 365 ¼ days. This revolution or orbit about the sun determines the length of our year.

At the same time that the earth is going around the sun in a path or orbit, it is also turning around and around from west to east on its *axis*. This is an imaginary line that runs through the center of the earth from North Pole to South Pole.

As the earth moves in its orbit around the sun, turning on its axis as it does so, its axis is not straight up and down. It is tipped or inclined 23 ½ degrees from the *per-*

pendicular (an imaginary line straight up and down to the plane or path of orbit). The axis is always tipped in the same direction and always about the same amount.

Two things cause the seasons of the year. They are the movement of the earth around the sun, and the constant direction of the tilt of the axis. These two things cause the changes of spring, summer, fall, and winter.

If the earth's axis were at right angles to its path we would not have any seasons. Each day the sun would pass through the sky in the same path. Day and night would be

The Gemini 11 astronauts took this picture of the earth from 851 miles above Australia — the highest point ever reached by man.

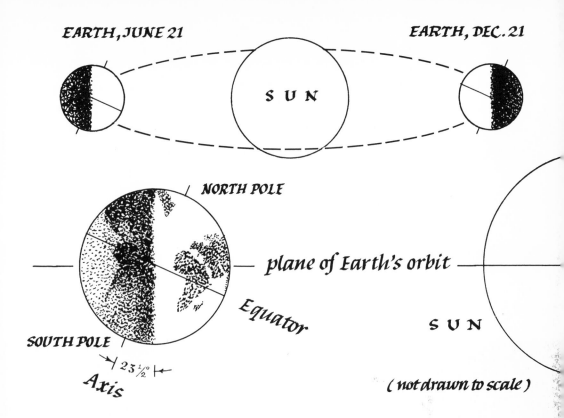

EARTH, JUNE 21

EARTH, DEC. 21

S U N

NORTH POLE

plane of Earth's orbit

Equator

SOUTH POLE

23½°

Axis

S U N

(not drawn to scale)

twelve hours long at all places on the earth and at all times of year. Weather would be the same every day except for random variations that had no definite plan.

If the axis were inclined but the earth did not revolve about the sun, the heating at each location would be the same for each day. This would be true because, although the days and nights would not be twelve hours long except at the equator, they would remain the same length at any one particular place throughout the year. There would be no change of season.

The change of season depends on both the tilt of the axis and the orbit around the sun.

The sun seems to move through the sky because the earth is rotating. As your part of the earth turns toward the sun, the sun appears to come up in the east. When your part of the earth turns away from the hot, bright sun, the sun seems to set in the west.

The earth always turns or rotates in the same direction. Day begins when your part of the earth turns toward the sun. As the part of the earth you live on turns away from the sun, it is night. When the earth once again turns around to the sun's light, it brings day. Like the tides, the rotation of the earth is continuous and endless.

To understand how rotation of the earth causes day and night, you can do a simple experiment.

Take a globe that shows all of the countries of the earth. If you do not have a globe, use an orange, a ball, or some round object. Place a small mark on one side of the globe to show where you live.

Shine a light on one side of the globe. You can use a flashlight or a lamp. The light will represent the sun. On the closest side to the light, or sun, it is day. On the side where the shadow falls, it is night.

To show how night changes to day, slowly turn the part of the globe where you live toward the light. As the part of the earth you live on turns away from the light, you can see day turn to night. As the place where you live again comes into the light, night turns to day.

You know that twenty-four hours go by between noon of one day and noon of the next as the earth rotates on its axis. One complete rotation takes twenty-four hours.

If you have a top, it will help you to understand how the earth rotates on its axis. Although the spinning top will not show the tilting of the earth on its axis, it will help to make clearer the rotation of the earth. The spindle (the iron or plastic point that goes through the center of the top) resembles the axis of the earth. The ends of the spindle, both

top and bottom, are like the points of the earth that we call the North Pole and the South Pole. As you watch the top spinning on its spindle, think of the earth making one complete turn on its axis each twenty-four hours.

While the earth is rotating on its axis, as well as making an orbit around the sun, the moon at the same time is going around the earth. The moon orbits around the earth once each lunar month, or about once each 29½ days.

A force, which you cannot see, called *gravity* holds the moon and the earth and the sun in their place in the universe. Gravity also holds the waters to the earth and keeps the seas from whirling off into space.

For these reasons gravity is one of the unseen forces of great importance to us. Without the force of gravity, the earth, the moon, and the other planets would fly off into space.

In order to understand tides, you need to know a bit about gravity. According to an old story, our learning about gravity was just a matter of chance. A good many years ago, on Christmas Day in 1642, Isaac Newton was born in the village of Woolsthorpe in Lincolnshire, England. He grew up to become one of the greatest scientists in history.

One day Isaac Newton was sitting beneath an apple tree in his garden, or so the story goes, when an apple hit him on the head as it fell to the ground. Newton wondered if the force that pulled the apple to the earth was the same force that held the earth and the other planets in their orbits around the sun. If the apple had not hit Isaac Newton on the head, we might never have known about gravity.

In 1687 Newton told the world about gravity. He gave us the *universal law of gravitation*. He explained in this law that two things are important. One is the mass or weight of objects, and the other the distances between them. Both of these make differences in the force of gravity.

Have you ever had a tug of war with your friends pulling on the two ends of a rope? Gravity acts very much the same. The largest and sometimes strongest objects have the greatest pull, except that the *distance* between two objects is important in the pull of gravity.

Another example that is easily seen is the pull of a magnet. A good experiment is shown in the illustration to the left. Here a string is tied to a large nail. The string is thumbtacked to a board. Then the magnet is held above the nail. If you own a magnet, experiment to see how far away from the nail you can hold the magnet and still have the nail stay in the air.

This is about the way that gravity works, and you can see that the distance between objects influences the pull of gravity. This is true, too, of the force of gravity in relationship to the sun and the moon and the earth.

If you take a strong piece of string and attach it to a large heavy ball, and if you whirl this ball around in a circle, you will have a good example of orbits of planets.

As you hold the end of the string, you must pull hard to keep the ball "in orbit." The heavier the ball and the faster the orbit, the more you have to pull on the string. Then let go of the string and see the result of "gravity" stopping!

Isaac Newton explained that the mass or weight of the sun is so much greater than the earth that the size of the sun makes it possible to pull the earth toward it. The sun is more than a hundred times larger in *diameter*, or the distance through the center. It is a million times greater than the earth in volume. Although the sun is 93 million miles away from the earth, it exerts a great force because of its mass and pulls the earth toward it.

Look at the globe you have been using in the experiment of rotation of the earth. Imagine another globe one hundred times larger. This will give you an idea of how much larger than the earth the sun really is.

At the same time that the sun is pulling the earth, the stars are pulling the earth as well. The stars are so far away that their pull of gravity is very small; in fact we do not notice their pull at all.

We do feel the gravity from the moon, however. Although the moon is not as large as the sun or the earth, the moon is so close to the earth (in comparison with the sun) that it has a great gravitational pull on the earth. The moon is only 238,857 miles away from the earth. Think how much nearer the moon is to the earth than is the sun. Compare 238,857 miles·with 93,000,000. The difference in distance is really a tremendous one.

Now with this information about gravity, you will be able to understand more clearly about the moon's influence on tides.

CHAPTER III

What Causes the Tides?

GRAVITATIONAL ATTRACTION CAUSES TIDES. Because water, which covers about three fourths of the earth's surface, is a liquid and free to move around, it shows the effect of the moon's gravitational pull upon the earth.

The moon's gravitational pull lifts the sea like a wave. Actually, that is what a tide really is. The rise of the water is slow because it is a very long moving wave. The wave's motion reaches to the bottom of the sea.

The moon pulls hardest on the part of the earth nearest it. The pull lifts the water on that part of the earth. The water bulges out toward the moon. Where the bulge is greatest, it is known as high tide. When the moon is on the horizon, and the bulge is the lowest, it is low or ebb tide.

At the same time that the moon lifts the water up on the part of the earth nearest it, the moon pulls *least* on the

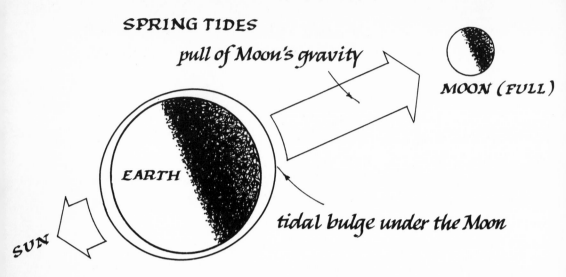

SPRING TIDES

pull of Moon's gravity

MOON (FULL)

EARTH

SUN

tidal bulge under the Moon

opposite side of the earth. The water on the opposite part of the earth bulges *away* from the moon, because of the lessening of the pull.

Sometimes this is explained by thinking of it as a "stretching" along the axis. The moon stretches the water along one part of the axis, and the sun stretches it along the other part. This tends to decrease the elongation or to lengthen the tidal bulge. Where there is a twice-daily, or semidiurnal tide, the rotation of the earth and the result of the sun and moon's pulling forces causes high water on both side of the earth.

The earth and moon pull on each other. At the same time they revolve about a common center, the sun. The outward force of the earth and the moon just balances the inward pull, and so they continue in their orbit around the sun.

As the moon circles the earth, it pulls the waters on different parts of the earth into high tides. These high tides, as you know, occur fifty minutes later each day. This takes place because the moon rises fifty minutes later each night. The moon rotates about the earth in approximately twenty-nine days and so is about fifty minutes later each day in reaching the same position.

Sometimes the local tides may follow this cycle or orbit, but they may be slower because of local tidal features. In some places (as in the Gulf of Mexico) there is only one low and one high tide each twenty-four hours and fifty minutes. This is called a *diurnal*, or daily, tide.

The earth constantly spins around, bringing another part of the globe beneath the moon. Each part of the water of the earth is pulled toward the moon. This bulge toward the moon, or high tide, moves around the earth as the earth continues turning.

There are times when the sun and the moon both pull in a straight line, and then the tides are *very* high. People who study the tides know that the highest ones of the month occur when the moon is full and again two weeks later at the new moon. At these times, the sun and the moon are in a straight line with the earth.

When the moon and sun are in line with the earth

and pulling together, *spring* tides, or tides of higher than average range, occur. The word *spring* has nothing to do with the season of the year. It comes from the German word for *jump*. Spring tides occur twice a month all year long, not only during the spring of the year. During these spring tides, the sun adds its gravitational pull to that of the moon's. Then the water washes up to a high point on the shore.

When the sun and the moon are pulling at right angles to the earth, the tide is lower. This happens at the first and the third quarters of the moon. When you can see a quarter of the moon, the earth and the moon form the three points of a triangle with the sun. The moon and the sun pull against each other. (Think again of the moon as stretching along one axis and the sun along the other.)

When the sun and the moon are pulling on the water of the earth at right angles during the first and third quarters, the high tides are only moderately high. These are often called tides of *minimum range* or *neap* tides. To call a neap tide a low or ebb tide is not correct. The exact meaning of a neap tide is quite different. A neap tide is the level at which the high water is the lowest. In other words, it is the lowest high tide!

This is difficult to understand, but tides are not simple things. In fact they are so complicated that men invented tide-predicting machines to help forecast tides. These tide-predicting machines have been in use for almost ninety-five

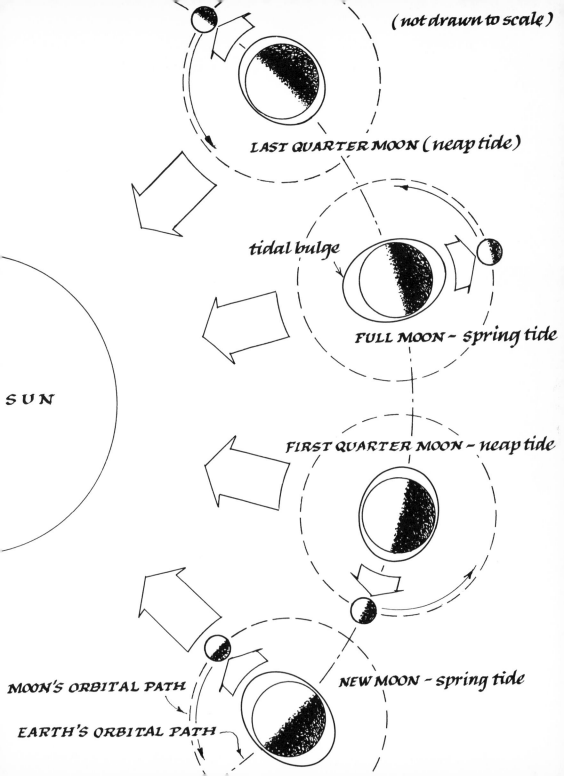

(not drawn to scale)

LAST QUARTER MOON (neap tide)

tidal bulge

FULL MOON - spring tide

SUN

FIRST QUARTER MOON - neap tide

MOON'S ORBITAL PATH

NEW MOON - spring tide

EARTH'S ORBITAL PATH

years to help men know quite accurately when spring tides, neap tides, and ebb tides are to occur.

William Thomson, who was later knighted and then raised to the peerage as Lord Kelvin, invented the first tide-predicting machine in 1872. Lord Kelvin has sometimes been called the first electrical engineer. His tide-predicting machine drew a line picture of the curve of the tide.

Approximately ten years later, William Ferrel, a member of the United States Coast and Geodetic Survey, designed a machine that was used by the Survey for over thirty years. On the face of this machine were shown the times and the heights of the tides.

Approximately twenty-five years ago, the Survey completed and put into use a simpler machine that pictures the curve of the tides and shows the times and the heights as well. The tide tables this machine predicts are published today by the survey.

In 1919 a German machine was invented that could predict a whole year of tide tables in one day. The latest and most complicated tide machine is the one at the German Hydrographic Institute in Hamburg.

However, no matter how accurate the machines may be, other things affect tides and make even the best tide table sometimes incorrect. Tides as shown in tide books are not always correct to the exact minute or to the stated height. They predict the sea level as it would be if only the sun and the moon acted on the water. Winds and heavy storms farther out at sea or on shore are not considered in the listings in the tide books. These could change the tides as shown in the tide tables.

It is quite impossible to forecast months ahead how strong the winds may be on a specific date. It is also difficult to tell the amount of sand that may be piled up by the waves from day to day. These things often influence the water level and cause the predictions of men and machines to be inaccurate, although large variations in tides can be predicted. Forecasts, even if they are not exact, give approximate high and low waters. They are of great help to many men who work or live near the sea.

Variations in tides can be anticipated by tide-predicting machines.

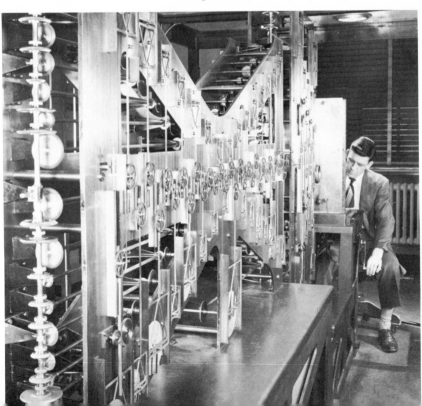

A high and low water level caused by a bore at Cooks Inlet, Alaska.

CHAPTER IV

Ranges and Differences in Tides

THE MORE YOU KNOW ABOUT TIDES the more interesting they become. Tides are different in various parts of the world. Even along the coasts of our continent there are wide differences in the ranges of the tides.

Simply stated, the difference in tide level is caused by the power of the wave motion in relation to the size and depth of the ocean basin and the shape of the coast.

In other words, tides are influenced by *sizes*, *depths*, and *shapes* of ocean basins. These basins are pool-like places on the ocean floor that vary in size and in depth. In each basin the waters tend to slosh back and forth like water in a pail that is being carried. These sloshings sometimes work for or against the tide.

Another surprising fact about tides is that they turn quicker and with more force at the narrow openings of creeks and other inlets than they do on nearby open sea-coasts. If you are near the opening of a creek or river at the turn of the tide, use great caution. The water confined in a narrow space rushes to get out. Do not be caught by a tide that could sweep you out to sea.

Tides do not come in and go out at the same time on any given day along the coastlines of our country. The range of tides, because of the differences in coastlines, often varies from one coastal town to another, even within a few miles.

On the Atlantic coast driving north from Florida you might leave in the morning with the water at high tide. It could be high tide again a hundred miles north by the time you got there which, of course, would be almost two hours later. It is a known fact that tides on any particular day come in later along the Atlantic coast as you go north toward Nova Scotia.

The range of the tides on the Atlantic coast includes two high tides of nearly equal levels each day, and two almost equal low tides.

On the Pacific coast, however, tides are quite different. They are commonly called *mixed tides*. Often the words *predominately semidiurnal* are added to show there are twice-daily tides. Each twenty-four hours on the Pacific coast there is a high-high tide and a low-high tide. There is a low-low tide, and there is a high-low tide. Anyone wanting to go to the beach on a low tide usually goes at a low-low tide. Then the water is lowest of all.

Along the coasts of our country, all newspapers pub-
lish tide tables to show when low tides and high tides will
occur. In many towns and various areas this information is
put into small tide books or pamphlets that you can carry
with you when you go to the beach. Here is a page from a
Pacific coast tide book:

TIDES OF ESTERO BAY REFERENCE STATION

SEPTEMBER

	Day	LOW TIDES				HIGH TIDES			
		A.M.	ft.	P.M.	ft.	A.M.	ft.	P.M.	ft.
Sunday	1	3:38	−0.4	3:09	1.9	10:04	4.1	9:15	5.9
Monday	2	4:09	−0.6	3:48	1.5	10:30	4.4	9:55	6.0
Tuesday	3	4:42	−0.6	4:30	1.2	10:59	4.6	10:35	6.0

Note: Tides in San Luis Bay are minus five minutes and for San Simeon
  Bay plus five minutes.

The first time you use such a table you will need some
help in understanding it. For example, this tide table is for
Estero Bay, along the central coast of California. If you
were going out to walk on the beach at low-low tide along
the bay, you would go very early in the morning on Sunday, September 1. On this date you would have a *minus*
tide. This would be a water level below that of the low-tide
level. The minus tide on Sunday is −0.4. On Monday and
Tuesday of that week it is −0.6, an even lower tide.

The note under the tide table for the first three days
of September of this certain year tells you that the tides in

San Luis Bay, south of Estero Bay, will be five minutes earlier (or minus five minutes) from 3:38 a.m. The low tide will be at 3:33 a.m. at San Luis Bay. At San Simeon Bay, north of Estero Bay, the low tide in the early morning will be five minutes later or plus five minutes. This would make the low tide occur at 3:43 a.m.

Now compare the Estero Bay Reference Station tide table with one from Sanibel Island, Florida, in the Gulf of Mexico. Are they alike or different? Which table gives you more information?

SANIBEL ISLAND, FLORIDA

NOVEMBER

		LOW TIDES		HIGH TIDES	
Day		A.M.	P.M.	A.M.	P.M.
Saturday	2	6:51*	6:59	12:09x	1:24*
Sunday	3	7:33x	7:32	12:40x	2:11*
Monday	4	8:16*	8:08	1:14x	2:59

* Strong   x Very Strong

Tides vary in other parts of the world, too. In the Gulf of Mexico, the range of the tides is not more than one or two feet. On the island of Tahiti, in the South Pacific, the difference in tides is less than twelve inches. Yet along the coasts of Britain the differences between high water and low water are commonly between twenty and thirty feet. Again no one reason can explain this, but the shape of the land, currents, and their effects on tides do influence the amount of rise and fall of tides.

The range of tides on the Atlantic coast of our country is generally three to four feet. There are variations in different areas. For example, along the coast of the island of Nantucket, off the Massachusetts shore, the tide lifts only a foot.

On the Pacific coast, the range of the tide is from four to eight feet. This will vary from area to area, too. Look once again at the tide table for Estero Bay. It shows for the three days listed a variation in height of tides, but they fall within the average of four to eight feet.

An eight-foot tide range covers quite a horizontal distance on a sandy beach, especially if the beach has a long, shallow rise.

There are places in the world where tides of twenty to thirty feet have been recorded. In some places it has been reported that tides have climbed more than fifty feet from low tides to high tides.

In harbors, where the rise and fall of the tide is quite noticeable, gauges are built to assist seagoing men. A tide gauge is like a thermometer. Instead of showing changes of temperature, tide gauges show changes in depth of water. They are of great help to men in piloting ships in and out of harbors.

On the banks of rivers that empty into harbors where the tide comes in and goes out, wide boards are often placed vertically along the sides of banks or cliffs or on piling of wharves. On these boards are painted numbers to show the depth of the water.

Along the Seine River in Paris, one tide gauge is painted on a rock wall that protects the banks of the river. In bays and rivers in America you will see tide gauges, too. These will be marked to show the depth of the water above the floor of the bay or river.

Since these markers on rivers are affected by splashing water or wash from the ships as they come and go, often a different type of gauge is used to show the tidal range. On seacoasts as well, measurements must be taken in such a way as to avoid the effect of the waves if an accurate measurement is needed. The illustration shows how a tank is placed on the shore above the high-water mark. The bottom of the tank is three to six feet below the lowest low water. A flexible pipe leads to the tank from the river or ocean. The end of the pipe is covered like a sprinkler on a watering can to keep sand or silt from entering the two- to three-inch pipe.

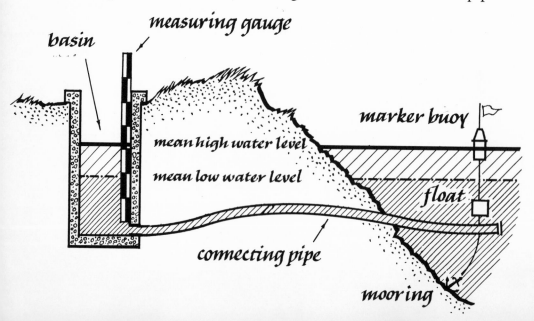

basin

measuring gauge

marker buoy

mean high water level

mean low water level

float

connecting pipe

mooring

To keep the pipe from becoming embedded in the sand or silt, it is held off the floor of the river or ocean by a buoy. The level in the tank is unaffected by waves and thus an accurate measurement of the depth can be shown on a marked pole.

One of the most famous ranges of tides in the world is on our own continent in the Bay of Fundy. It supposedly has the highest tides in the world. Twice a day, 3,680 billion cubic feet of water rush in and out of the mouth of the bay. Through this funnel-like entrance surge giant fifty-foot tides known the world over.

The Bay of Fundy is actually a huge arm of the Atlantic Ocean that extends into the land, with Maine and New Brunswick on one side and Nova Scotia on the other. The Bay of Fundy ends in two smaller bays, the Chignecto Bay and the Basin of Midas. Because of the narrow and shallow bay, large masses of water crowd into this small area. This contributes to the high water in the Bay of Fundy as well.

You can actually see what happens in the Bay of Fundy by using your own bathtub. Put some water into the bathtub (or you can use a large flat pan of water instead). Take a board or small paddle and push the water to the end of the tub or pan. The water will slosh back and forth with a definite rhythm. If you continue to push with your paddle with the same rhythm, the sloshing gets higher and higher with each push of the paddle. This sloshing is similar to the action of the high water in the Bay of Fundy, where the water rushes into the narrow entrance and then sloshes back and forth until it returns to the sea with the outgoing tide.

The rivers emptying into the Bay of Fundy are an interesting part of the tide range, too. The town of Saint John is situated at the mouth of the Saint John River. There is a narrow rocky gorge which constricts the stream of water where it empties into the northwestern side of the Bay of Fundy. At low tide, the river falls about fifteen feet into the harbor. At high tide the flow is reversed and a cataract roars upstream. The force of the tide is so great that it is felt

The difference between high [above] and low [below] tides at the Bay of Fundy is the largest tidal range in the world.

seventy-five miles up the river. Near half tide, or between high and low tides, the river is level and navigable.

Moncton is eighty-five miles northeast up the coast from Saint John on the Petitcodiac River, which empties

into Chignecto Bay at the end of the Bay of Fundy. The tide sweeps up the Petitcodiac River to Moncton in the form of a great wall of rushing water called a *bore* or *eagre*. It is often correctly called a *tidal wave*, because it is caused by the incoming tide.

Some authorities say the word *bore*, meaning hole, is of Norse origin. The word *eagre* or *aegir* means *sea-god*. Other authorities say the word *bore* comes from the old English use of the word meaning *wave* or *carry*.

Bores occur where a sand bar or some other obstruction at the mouth of a river holds back the rising tide. When entering these rivers the tidal wave often rises so rapidly that it makes a wall of water. The tide piles up higher and higher. Finally it is big enough to flow over the obstruction. Then it rushes forward as several waves, one behind the other, or as a single great wave like a solid wall of water.

When the bore comes into the river, it is a thrilling sight to see from shore but it is a dangerous thing to meet in a small boat. Men of the sea on the Bay of Fundy pay a great deal of attention to the tide tables. They listen to the weather reports and watch for signs of heavy winds. They are most cautious if heavy winds occur during the time of spring or higher than average tides, for this combination of heavy winds and high tides will cause the bore to rise to its greatest height. This could bring the bore at Moncton to a height of six feet.

Such variation in tides makes a great difference in the

lives of the people who live nearby. In Saint John, when the tide is low, the fishing boats lie on their sides in the mud. High wharf structures at low tides are many feet above the small stream of water and the mud flats below. At high tide the bay becomes very deep. The fishing boats turn with the tide and soon their masts are erect and straight.

Bores occur in other places in the world as well. One of the more famous outside our own continent occurs in southwestern China. Hangchow Bay, at the mouth of the Tsientang River, is nearly one hundred miles long. The funnel shape of the bay with a narrow entrance to the river causes a bore when the tide, rushing into the river, rises in a wall of water. This may range from a few feet to fifteen feet high. In flood season it may rise as much as thirty feet and move at almost fifteen miles an hour.

Another spectacular wall of water is the pororoca bore at the mouth of the Amazon River, in South America. It is not as large as the Hangchow bore, but even then it moves up the river almost two miles before it flattens out. It rushes up the river at fourteen miles an hour and is like a waterfall almost sixteen feet high. The roar of the water can be heard for miles.

In many areas of the world there are great ranges in tides without bores. These occur in places where there are wider entrances to the harbors. In these areas the tidal waters can flow over a large area. They do not have any constriction that holds the waters back.

In Korea at the port of Inchon the difference between high tide and low tide is twenty-seven feet, one of the greatest in the world. The tide at Inchon runs so swiftly that a large ocean-going ship is completely turned around at anchor in less than twenty minutes.

Ships must anchor outside the harbor at Inchon. Large barges come out to meet the ships. All cargo must be piled onto these barges to transport the supplies to shore. Barges can go into the port at high tide. At low tide the whole harbor is nothing but mud flats. Even at high tide, although there is a twenty-seven foot tide, the tide spreads out over such a large area that the water in the actual harbor is not deep enough for large ships.

During the war in Korea, the United States Armed Forces had to know the tide tables and the high- and low-water lines in planning the landing of troops in Inchon. They planned the land attack weeks ahead so that the tides could help them in landing troops. The Armed Forces needed a high tide to get the landing craft and other small ships carrying troops over the miles of mud flats.

In Europe off the coast of France, the famous small island of Mont-Saint-Michel can be easily reached by road from the mainland at low tide. At high tide the waters swirl around it forty feet deep. Years ago people who lived on the island could only get to shore at low tide across mud flats. Today there is a causeway that is built above the high-water level.

Mont-Saint-Michel, a small island off the coast of France, is a place often visited by tourists. From the causeway, visitors can watch the tides come in and go out around the island that has been a famous place since the twelfth century.

A sister island, Britain's Saint Michael's Mount, stands across the English Channel. The causeway to the island, which connects it to the Cornish shore at Marazion, is only above water at low tide. Flood tide sweeps in so swiftly that visitors to the island must run through water to get back to shore if they do not cross on the low tide. Island children cross the bay to school. Sometimes heavy seas isolate their island home.

There are many other places around the world that have a tidal range of thirty feet. Among them are Puerto Gallegos in Argentina, Cook Inlet in Alaska, and, in Canada, Frobisher Bay on Davis Strait and the Koksoak River emptying into Hudson Strait. As a rule strong tidal waters occur in straits, tidal rivers, over shoals, and off capes.

One of the greatest differences in tide range within a short distance occurs on the eastern and western ends of the Panama Canal. Colón, on the Atlantic side, has a tide rise of only one foot. It has only one high and one low tide daily. At Balboa, on the Pacific end of the Canal, only thirty miles away, the tide rises fourteen feet. It has two high and two low tides each twenty-four hours and fifty minutes.

A look at the map will help you to understand some of the reasons for this. The shape of the land and the water on the Atlantic side is quite different from the Pacific side. The depth of the water and the basic tide pattern accounts for other differences in the tide range.

Large ocean-going ships like the "Queen Elizabeth"

wait for slack water to come into New York Harbor. Otherwise, the tidal waters with their strong force could make docking at the pier more difficult. Tidal waters could sweep the ship against the pier with force strong enough to do great damage to even a huge vessel.

This is not only true of the Hudson River, which flows into New York Harbor, but it is true of other rivers as well. The Sacramento and the San Joaquin Rivers, which enter San Francisco Bay, and the Columbia River, which flows into the Pacific Ocean at Astoria on the Oregon coast, are good examples of West Coast rivers where strong tidal waters occur.

When the tidal currents pass through a strait, such as a narrow inlet into a bay, or between an island and the mainland, the speed of the current is very rapid. The tidal current in the narrow passage from the East River into Long Island Sound is about five to six miles an hour. In the English Channel, between England and continental Europe, tidal currents as fast as nine miles an hours are known.

Tidal waters often run up rivers to a point many feet above sea level and many miles from the harbor opening. The tide runs 150 miles up the Hudson River to Troy, which is five feet above sea level. The tidal range here is more than two feet. There is even a tide at Montreal, 280 miles up the Saint Lawrence River. All of these tidal waters occur against the downward flow of the rivers.

CHAPTER V

Waves on the Shore and at Sea

WHEN YOU LOOK AT THE WAVES ROLLING IN, it seems as if the water is actually traveling in toward land. However, it is more complicated than that.

When a *wind wave* first starts far out on the ocean, the water is hit by a breeze or wind. Sometimes this causes a ripple or a roughening of the surface of the water. If the wind is heavy enough, this ripple will change into a small wave. If the wind is strong, the small wave becomes a larger wave.

If you have been on a ship at sea or in a rowboat in a harbor, you may remember a day when the water was very calm. You may have seen the start of a wind wave as you watched a ripple become a small whitecap.

Wave formation is rarely this simple, however; it is more than just one wave coming in after another. Often little ripples will be seen on larger waves. Many times ripples ride in between waves. Smaller waves often ride on the backs of larger ones.

The little ripples are very important to the starting of waves. The ruffling of the water makes it easier for the wind to catch hold of the ripples. More of the energy in the wind pushes the little ripples and waves. This makes the wave formation even more uneven than it would be without the wind.

If you take a rope and tie it to something solid, and then give a twitch to the rope, you will see a good example of how a wave works. The "waves," or up-and-down motions of the rope, follow one after the other along the rope to where it is tied. No part of the rope actually moves forward. This is exactly what happens with waves in the sea.

You can see the same thing in the fluttering of a flag in the breeze. The ripples or "waves" in the flag caused by the wind are similar to the waves on the ocean. Think of the rope twitching or the flag flying in the breeze the next time you look at waves on the ocean or study pictures of waves in a book.

You can see this quite plainly when the sea is calm and the water rises and falls in long, smooth ridges that may have started far away.

Have you seen a field of grain or tall grass on a windy day? The tips of the grass are carried ahead or bent down, but when the wind stops, they return to their original position. This happens with the waves in the sea, too.

Another way that you can tell that waves themselves do not move is to watch a piece of driftwood several feet from shore and beyond the breaker line, bobbing up and down on a wave. Often it stays in the same place for long periods of time.

The up-and-down motion of waves is shown as well by small boats. Watch one and you will see that the boat merely rises and falls while the wave motion moves on.

All water is made up of tiny, minute parts called particles. Water in waves is made up of particles, too. They flow in almost a circle as the motion of the water goes up and down. Except in very shallow water, the circular motion of the particles is a constant part of each wave. Each time the particles go around in a circle, they go ahead slightly in the direction that the waves move.

Follow the arrows in this picture. Each arrow indicates how the tiny particles (shown by dots) go around in a circle.

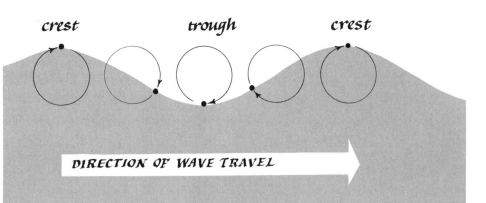

crest   trough   crest

DIRECTION OF WAVE TRAVEL

Over a hundred years ago, in the year 1802, Franz Gerstner of Czechoslovakia suggested that particles of water moved in a circle. At the *crest*, or top of the wave, he believed the water moved in the direction of the wave. Then it moved, just as a wheel turns, down into the *trough* (or lowest part of the wave), around, and up again to the crest of the next wave.

Gerstner thought it was the force or the motion of the wave that traveled across the sea. He thought that the water did not actually move ahead. We still think that this is true today.

As a mathematician, Gerstner was interested in numbers. In his studies of waves, he learned how they were formed and how they behaved. He used his knowledge of mathematics to change his findings into formulas. His work so many years ago was the beginning of many studies about waves.

Gerstner's model wave is never found in the sea. The ordinary waves, as you know, are very much more complex. Waves do not have a set pattern. This lack of a routine pattern, sometimes called the randomness of waves, is a very important part of knowing about them. Various sizes and shapes in no special order are all a part of the waves of the sea.

Today men are working in greater detail to understand waves, for there are still some things that we do not know about them. We do not have a complete answer to all the ways by which the wind sends its energy to the water to form waves. We do know that the individual particles of water are locked together loosely. These particles cling together and roll over each other, pulling them along.

The interesting thing is that each circle of the particles of water gets bigger and bigger as the wind blows more energy into the wave. This means that the wave motion gets larger. In a storm, the wind brings so much energy that the waves may become giant ones that reach many feet into the air.

The height and length of waves, and the time or period between waves, depend on several things. The speed of the wind and the length of time it has blown are important. The *fetch*, or how far it has blown without changing direction, also affects the height, the length, and the period or time between waves.

The length of a wave is measured from one crest to another. The height of a wave is the distance from the trough to the crest.

The time between wind-wave periods is the number of seconds between crests. (See A and B, page 48.) The time ranges between two and thirty seconds. (See X.)

A wave will rise as high in feet as one half the speed of the wind in miles per hour. If the wind is blowing ten miles per hour, the waves may be five feet high. However, this is only approximate. Too many other factors must be considered to give an exact figure.

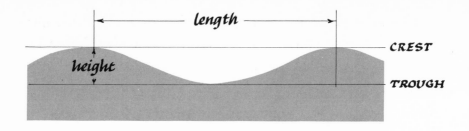

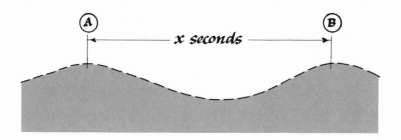

Next time you hear a broadcast over the radio or television that gives the miles per hour that the wind is blowing along the coast, think what it will mean to the waves along the shore.

Some scientists believe that in shallow water the up-and-down motion decreases more quickly than the to-and-fro movement. They think the orbit of the particles becomes an elliptical or flattened circle shape, rather than a true circle. It is thought that the forward motion of the orbit of the particles is most rapid in the crest of the wave, the backward motion most rapid in the trough. If there is a wind blowing, this would cause an increase in forward motion of the water.

Look at these illustrations. They show how the circle of particles becomes a flattened circle shape in more shallow water. This is because water along the bottom can only move horizontally.

In very deep water, the orbits are nearly circular, but the circles are smaller at greater depths. In more shallow water, the circles are flatter but are smaller nearer the bottom.

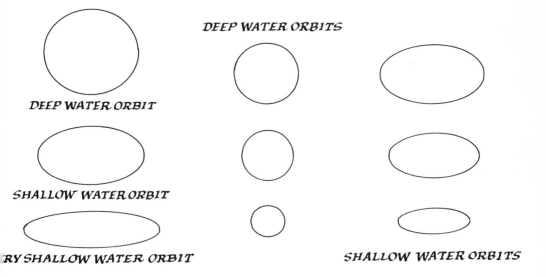

DEEP WATER ORBITS

DEEP WATER ORBIT

SHALLOW WATER ORBIT

RY SHALLOW WATER ORBIT

SHALLOW WATER ORBITS

When waves approach a gently sloping shore, the period of time of the wave remains unchanged, but the wave becomes higher and shorter. The crest of the wave arches forward. Because of lack of water in the front, the wave becomes steeper in front than the back. This causes the wave to rush forward and break on the beach to form *surf*.

An individual breaking wave is called a *breaker*. The height of the wave determines the place where it will break. Remember that a wave extends as much below the surface of the water as it does above. A high wave will break much farther out than a low wave.

When the heavy winds or gales lessen, the water cannot immediately make smaller waves. It is like a swing going back and forth. The swing does not stop at once even when someone stops pushing it. In a similar way, the force of the water still goes on to make up-and-down motions long after the winds die down.

This type of movement we call a *swell*. It is sometimes called a *ground swell* when it reaches quite shallow water. A swell is a long, smooth, and regular wave. It is a gradual rising and falling of wave motion. It is produced by winds and storms that usually start at a distant point from where you see the swell. Because it has started a long way off, the swell has lost much of its energy and comes smoothly and evenly toward shore.

Without the wind, the swells settle into a regular pattern. The water usually rises and falls almost the same way

Hawaiian surf riders glide down from the advancing front of a huge wave near Waianae, Oahu. This wave was estimated to be about nineteen feet in height. An increasingly popular sport, surf riding requires skill and coordination.

until the waves die out or reach shallow water near land. Such waves or swells travel long distances. Some of them have been found to have traveled more than a thousand miles.

Surf riders in La Jolla, California, know that the best surfing waves come in summer. Scientists tell us that these waves come over miles of sea from near the "roaring forties" in the Southern Hemisphere. Our summer season is winter time there, and the winter waves from far away make fine summer surfing in southern California.

A fact that many people do not realize is that it is not the direct push of the wind against the backs of waves that cause all of them to advance. Once a wave that is started by a back-and-forth action, or *oscillation*, has been set in motion, it will continue to run across the surface of even a calm sea.

The bow of a ship pushing ahead across the bay causes waves to lap against the shore long after the ship has passed. In very foggy weather, a watcher on shore knows a ship is passing off shore when waves that the vessel has started break on the beach.

When you throw a rock into a pond or river, ripples or waves start as soon as the rock hits the water. They continue even when there is no wind. Waves are formed in several ways and many things influence them.

One of the most destructive ocean waves has nothing to do with the wind. Although this wave is often *misnamed* a "tidal wave," because it looks like a rapid rise of the tide, it has no connection with the tide. This wave is caused by earthquakes and is called a *seismic sea wave*, named from the Greek word *seismos*, which means earthquake. These huge waves are also called *tsunamis* (tsū-nah'-mēs) from the Japanese term for "large waves in harbors."

In the earthquake in Alaska on March 27, 1964, the town of Seward was hit by thirty-foot-high seismic sea waves. Docks, warehouses, and railroad yards were completely destroyed. Freight cars rolled over and over, carried high off the ground by the huge tsunamis. Many of the

[Opposite] The USS ROWAN creates waves which rock smaller boats in the area long after the destroyer has passed.

high waves were covered with burning oil from storage tanks that were set on fire by the earthquake shocks. The flames made the waves even more terrifying as they came roaring in toward shore at a speed of more than 100 miles an hour.

The start of these huge seismic sea waves occurred somewhere off the southern coast of Alaska. Some eighteen miles down, the sea bottom heaved and plunged violently. This set millions of tons of water in motion. It was this motion of a tsunami or seismic sea wave that rushed to shore with tremendous force, causing great damage.

As these huge sea waves were hitting the coast of Alaska, the Pacific coast of our country and the islands of the State of Hawaii were lashed by huge waves.

The shape of the coast rather than the distance from Alaska determined the wave height that caused destruction miles away. For example, in Crescent City, California, 2,000 miles from Alaska, nine-foot waves swept the entire water-front of this coastal town, doing great damage. Yet in Puget Sound, 400 miles nearer Alaska but much more protected from the open sea, the waves rose only half as high.

One of the characteristics of a tsunami is that the huge wave rolls in from the sea over the land, then rushes back to the sea, taking houses, boats, and people with it. For a short time afterward, often as long as fifteen minutes, a quiet calm occurs. Families who live near the sea notice that the breakers do not come in on the shore. The sea water during this

As a result of the March 1964 earthquake in Alaska, a huge seismic sea wave caused tremendous destruction along the coast.

Destruction caused by a tsunami is almost unbelievable. In Hawaii in 1946, piers were washed away, ships damaged, homes destroyed, and many lives were lost [below].

quiet time is far out away from the shore. Reefs and rocks along the shore are uncovered as though it were a very low tide. Then the wave returns, carrying everything in its path far onto the land. Houses have been known to be moved eight blocks inland and boats and larger ships at anchor in harbors have been swept away many blocks from the water.

Often a third wave follows the first two. This happened at Kaguyak in Alaska in 1964. To escape the tsunami the people of the town had run from their homes on the waterfront to higher ground. Here they waited for the second wave. After it had passed, several men thought it safe to return to town for supplies. In rushed the third wave and swept some of the men along with houses and other buildings out to sea.

Eighteen years earlier, in 1946, seismic waves started near the Aleutian Islands. These waters traveled 2,000 miles southward at nearly 450 miles an hour. It took the waves less than five hours to reach the Hawaiian chain of islands. Along Pacific shores, they were recorded as far south as Valparaiso, Chile. The waves that reached Valparaiso took eighteen hours to travel the distance of 8,000 miles.

During the time the waves were traveling toward Hawaii, they were long, low waves, never higher than two feet. However, they measured 90 miles from crest to crest.

When these tremendously long waves came near the shoreline of Hawaii, they began to pile up. As each long, low wave came in, it pushed further inland. As the wave swept back to sea, it took houses and other buildings, automobiles, trees, people, and huge volcanic rocks with it. Men who have seen the tidal bore at Moncton, New Brunswick, say a tsunami looks very much like a bore as the great wall of water advances across the land.

The 1946 tsunami in Hawaii was reported to have had more than seven huge waves, each one supposedly larger than the preceding one. Also in this tsunami a second, even larger, series of waves arrived from the west. (The first ones had come from the north.) Scientists suggested that the waves were reflected from an underwater cliff off Japan and then from one in Oceania to come in on the sheltered side of the main island of Hawaii.

Today the Hawaiian Islands have a network of stations with special instruments to help protect them from surprise tsunamis. The United States Coast and Geodetic Survey operates a new audible alarm at seismograph stations that can tell immediately when an earthquake has occurred.

Tiny earthquake shocks occur thousands of times a year. About five thousand of them are widely recorded each year. Perhaps a hundred of them are major earthquakes. Many of these occur under the ocean or in uninhabited places. Only a small percentage of earthquakes cause tsunamis. Scientists are beginning to recognize the waves before they hit by carefully tracing them. They can tell how fast the tsunamis are traveling by the water depth. In the deep ocean areas, waves may travel 400 to 500 miles an hour. In other places, for example in the Bering Sea or the Gulf of Mexico, they may travel less than 100 miles per hour. Trained observers check from station to station. They alert the areas that lie in the path of the seismic sea wave. Plans can then be made for people to leave the low-lying areas near the shoreline.

Many of the early books about the sea told of waves one hundred feet tall. They described how ships were tossed about like pieces of cork. They described sea dragons that were supposed to lash their tails and cause huge waves. Many of these stories were thought to be in large part only the imaginations of the sailors.

Waves over one hundred feet high have been reported in the open sea. But most waves authorities give seventy-five feet as the highest reliable estimate.

In recent years the Oceanographic Institution at Woods Hole, Massachusetts, has supported some of the stories about the one-hundred-foot waves. The scientists who have been making the examination of facts believe that a large number of waves, each traveling at a different speed and attaining a different height, travel in the same direction. As they continually get into and out of step with each other, they produce alternating groups of large and small waves.

By chance every now and then a large number of them get into step at the same place and an exceptionally high wave occurs. The wave usually lasts only a minute or two.

Ships now carry instruments that can measure accurately the height of waves. High waves cannot be predicted with any assurance, but at least some of the old sailors' tales have proved to be actually true.

Another kind of wave is caused by high winds blowing toward shore. They produce *storm waves* that are accompanied by an independent rise of the water level called a *storm tide*. If these waves occur at the time of a high tide, the water level becomes even higher. The entire island of Galveston, Texas, has been under water from such waves.

The storm of March 1962 along the Atlantic coast is a good example of what happens when gale-force winds come at the same time as a high tide. Storm waves and storm tides did one hundred million dollars damage to property along the Atlantic shoreline. Damage in Delaware was said to have amounted to forty million dollars. On the New Jersey coast the storm damage was originally estimated at thirty million dollars, but amounted to much more.

Long Beach Island, off the coast of New Jersey, probably experienced the greatest damage. There are approximately twenty small towns on this narrow island. All of them were badly damaged.

Near the town of Harvey Cedars, 86 of 500 homes were washed away and over 240 others had extensive damage. Wind and sand damage, as well as water destruction, were seen everywhere. Two-story houses were completely covered with sand, while others had the sand washed from

Storms and hurricanes often cause extensive damage to coastal towns. Long Beach Island, N. J., was cut in two by a storm in 1962. In 1965, Hurricane Betsy flooded Miami Beach, Fla. [below].

under them. In some places the level of the sand dropped twelve feet.

Merton Rice of Washington, D.C., gave an eyewitness account to the Philadelphia *Evening Bulletin* for March 8, 1962. After leaving his car at Manahawkin, across Barnegat Bay from the island, Rice walked several miles toward Harvey Cedars in an attempt to find his father. He reached a point about two thirds of a mile from Harvey Cedars. Here he found the island cut in two by such deep water that he could go no further. Rice estimated that the cut was at least a mile wide.

Across this cut his father, Eugene Rice, was alone in his house in Harvey Cedars. In a personal letter to friends he told how he had stood on a dune twenty feet above the sea. From this vantage point on his property, he looked down the beach and saw that about a quarter mile from his house the dune had given way. The water was rushing in, heading for the road that ran down the middle of the island, which was about three blocks wide at this point.

"When I saw the water," Eugene Rice wrote, "I knew that I would have trouble getting to Harvey Cedars. I waded to the car, and started off in the pouring rain. The trip to town usually is less than five minutes. It took me well over an hour. I had to stop every few minutes to move obstructions out of the way. Finally I reached the firehouse that had been opened as a disaster shelter. It was a night to remember! Working in shifts, we swept constantly to keep the water from the front of the firehouse. All night and most of the next morning helicopters came and took the women and children to safety. Each trip they brought milk and other supplies back to us. Finally it was decided to evacuate everyone. My turn came and I flew out. Quite by chance, I ran into my son who had been searching for me."

At Surf City, Acting Mayor Robert A. Evans reported that barriers and sea walls blocking the ocean gave way and water streamed across the sand to Barnegat Bay, making two islands of the one. A channel, thought to be at least four feet deep, separated the sections. The island was about a mile across at this point.

Because exceptionally high tides and gale winds of up to eighty miles an hour hit the coast at the same time, the water level was far greater than expected. The storm waves and the storm tide brought great destruction to the entire area.

Often the rise of water during storm tides is so rapid that there is little chance for people to escape. Storm waves cause great loss of life and property. They have been known to shatter houses on shore and send pieces of the houses fifty feet into the air.

Other times storm waves strike against lighthouses, completely surounding ones that stand one hundred fourteen feet above the water. Storm waves have been known to sweep them from their foundations.

The first man to measure the force of an ocean wave

was Thomas Stevenson, father of Robert Louis Stevenson. A well-known lighthouse engineer, in 1842 he developed an instrument known as a wave dynamometer. This measured the force or power of the wave. With this instrument he studied the waves that battered the coast of Scotland where he lived.

Stevenson reported that during a storm at Skerryvore Rocks in the Atlantic, the top force of waves that had an average height of ten feet was 3,041 pounds per square foot. For twenty-foot swells this force increased to 4,502 pounds per square foot. During strong gales when waves were in excess of twenty feet, the maximum force was 6,083 pounds per square feet. No wonder that lighthouses were swept from their foundations!

CHAPTER VI

Currents of the Ocean: the Rivers of the Sea

THE WORD CURRENT comes from the Latin word *currere*, which means "to run." And that is what the currents of the ocean actually do. They run from place to place in a nearly set pattern, just as the rivers run to the sea. It is for this reason that we call the currents of the ocean the rivers of the sea.

Some currents move rapidly; others run slowly. Some currents can be seen as you fly over them in an airplane. Others can be felt by ships as the currents move along, but they cannot be seen.

Today you may read in books and in articles about the sea the word *gyre*, for oceanographers often speak of a current as a *great gyre*. Gyre means to move in a whirling path, and most currents do just this. The patterns they follow are likely to be circular in shape, although the circle may be tremendous in size. Many offshoots of the larger currents tend to break up this over-all circular pattern.

Look at a flat map of the world or a globe. First of all, you should know that the waters of the oceans of the earth are really one, although they are divided by land and have been given different names.

The southern ocean, called the Antarctic, is found near the South Pole. You will find this ocean on the bottom or southern part of a map of the region, or on the lowest part of any globe.

Three arms of water stretch out from the Antarctic. They are long and rather irregular. These we call the Atlantic, the Pacific, and the Indian Oceans. The Atlantic and the Pacific Oceans reach northward across the equator on either side of North and South America until they join the Arctic Ocean at the North Pole. This you will see at the top or northern part of the map or globe. The Indian Ocean sweeps around Africa and is stopped by Asia to the north and Australia and the islands of the South Pacific on the east.

Within each ocean there are currents that are determined by the position of land masses. These land masses change the currents as they flow around the coasts. The currents are also changed by other currents that approach and pass them or join them.

You can learn a great deal about currents by using a large bowl of water and blowing across it. If you continue this for a minute or more (use fireplace bellows if you get

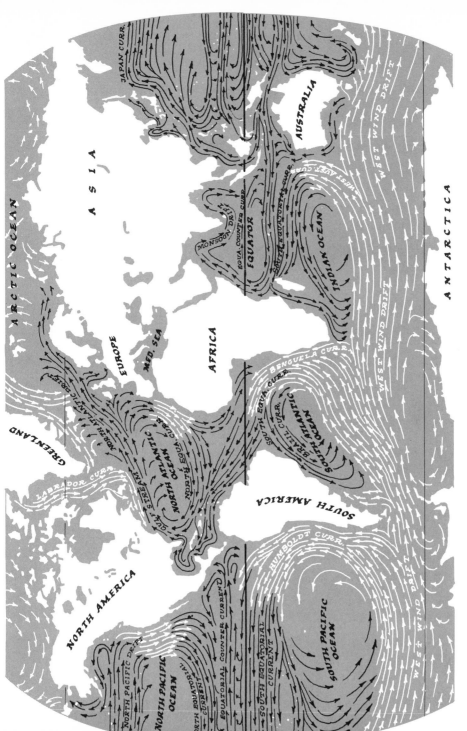

The position and direction of the major currents of the world.
The warm currents appear in black; cold currents in white.

tired of blowing!) you will see that the water moves from one side of the bowl to the other. With your blowing you have started a "current."

Scientists believe that ocean currents are created in the same way by the steady blowing of winds. The most important winds to influence currents are the ones that blow constantly and always in the same direction. These are found near the region of the equator.

Two things, the steady blowing of certain winds and the spinning of the earth on its axis, are the most important factors in shaping ocean currents.

The spinning of the earth has a curious effect on wind and water and actually on all moving objects. It causes a thrown ball, a boat, a ballistic missile, or other objects to turn slightly to the right in the Northern Hemisphere (above the equator) and to turn slightly to the left in the Southern Hemisphere (below the equator). Men who fire long-range missiles have learned they must make allowances for this if they want to hit their targets. This slight turning of objects to the right or to the left is called the "Coriolis effect." It was named for the French mathematician who first described it. It is often spoken of as Ferrel's law, because William Ferrel was one of the earliest scientific writers to develop it in connection with winds. He was not the first one to know of the law, however.

Trade winds are the steadiest winds. Often they are just called "the trades." The trade winds always come from

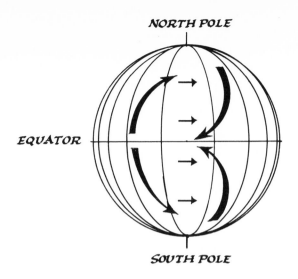

the east and blow toward the west. In the Northern Hemisphere the trade winds blow to the southwest, and in the Southern Hemisphere to the northwest.

While some writers say "to the left" and "to the right" in talking of the flowing of the currents and the blowing of the winds, others call it "clockwise" and "counterclockwise." This is an easy way to remember the Coriolis effect, for you can think of the hands of a clock or watch. Then you can remember right as being the way the hands turn, or clockwise, and left, or counterclockwise, as being in the opposite direction.

As the earth swings around the sun in orbit, it turns on its axis in such a way that the seas near the equator receive the direct rays of the sun. You can see this plainly if you look at a globe. You can see that the Antarctic Ocean at the South Pole and that the Arctic Ocean at the North Pole do not get as much heat from the sun.

The heating of the ocean water at the equator has another effect on the water. When the sun warms the water, the water expands and the water level actually rises a few inches. It is enough to produce a slight slope. As a result, there is a tendency for the warm water to run downhill toward each pole. The effect of this, however, is very small compared with that of wind on surface currents. Some authorities feel it may be important in exceedingly slow, deep circulation. Since water contracts and becomes heavier as it cools, the cold water from the poles sinks below the warmer water. This changing of warmer water and colder water is an important part of the ocean currents of the earth.

Under the constant driving of the trades, a warm current flows westward on each side of the equator. This westward flow forms the currents of the Atlantic.

One of these is the North Equatorial Current. It flows from east to west in the area of the northeast trade winds. It joins a branch of the South Equatorial Current. Some of the combined current enters the Caribbean Sea. Some of it passes outside the West Indian Islands.

What is probably the best-known current, the Gulf Stream, is really a continuation of these currents. It is one of the most interesting and most spectacular currents. From the air, the Gulf Stream of the Atlantic Ocean shows as a wide band of blue in the green water that surrounds it.

The Gulf Stream moves northward past the east coast of the United States. Some scientists who study currents

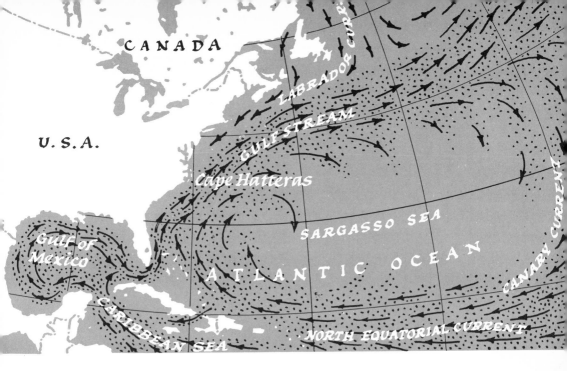

think of the Gulf Stream as starting at Cape Hatteras in North Carolina. South of this point, as it sweeps through the Gulf of Mexico, the Gulf Stream is often called the Florida Current. Other men call the current the Gulf Stream after it passes the narrow Straits of Florida.

North of Cape Hatteras the Gulf Stream is fifty miles wide in some places and has a depth of almost a mile. It moves about five miles an hour and carries along more than four billion tons of water a minute.

Benjamin Franklin drew the first map of the Gulf Stream. When he was Deputy Postmaster of the Colonies, he wondered why the English ships were so slow in getting to this country. He found the answer from a Nantucket whaler, Timothy Folger. Folger told Franklin that the American ships had learned to keep out of the Gulf Stream when they returned to this country. The English ships took two weeks longer, for their captains stubbornly went against the current. After Benjamin Franklin published the first chart of the Gulf Stream, he worked hard to show seagoing men how to find and use the strong current.

Benjamin Franklin's original map of the Gulf Stream.

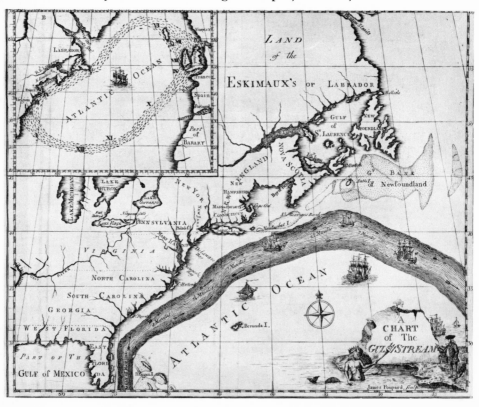

One of the things Benjamin Franklin suggested to the captains of the ships was to pick up buckets of water and chart the temperatures. When higher temperatures were listed, the captains could tell if they were near or in the Gulf Stream. Then each captain could plot a course to stay away from the current if it were flowing against him, or to stay with the Gulf Stream if it would help his ship to reach its destination faster.

Scientists tell us that part of the force of the Gulf Stream comes from the fact it is actually flowing downhill. Strong easterly winds pile up surface water in the narrow Yucatan Channel and in the Gulf of Mexico. The level in these areas becomes higher than in the Atlantic Ocean.

Off the southwest tip of the Florida Keys the sea is about seven and a half inches higher than at St. Augustine, farther up the Florida coast. The earth's rotation sends the water toward the right side of the current. Actually the Gulf Stream slopes upward toward the right. It is a known fact that along the coast of Cuba the ocean is about eighteen inches higher than along the mainland.

The Gulf Stream is a warm current. It has come from a tropical ocean where the temperature often is 80 degrees Fahrenheit. The Gulf Stream comes into the Atlantic Ocean bringing with it the warm water from many miles of traveling in warm climates.

As the Florida Current or the Gulf Stream reaches Cape Hatteras, it turns outward into the Atlantic. This is caused partly by the coastline of the United States, as well as by the Coriolis effect of the earth's turning.

Soon after the Gulf Stream leaves the eastern coast, the current slows down. It does not make more than ten or twelve miles a day. Scientists refer to it as an "eastern drift" and not a true current.

After many miles, at last one portion of the drift turns southward and becomes the Canaries Current. With the help of the northeast trade winds, this portion completes the circle tour of the North Atlantic as a cold current. As it joins the North Equatorial Current once more, it has lost its warmth after many miles of traveling with cold water.

The other portion of the drift with the help of the southwesterly winds continues north. This part of the drift, called the North Atlantic Drift, runs between the British Isles and Iceland. It sends small branches of the current into the shallow water near Great Britain. This warm current keeps the seas free of ice and brings warmer weather to all the lands it passes near.

Coming down from the lands nearer the North Pole is the Labrador Current or Arctic Current that brings cold water from this region. It starts from Davis Strait and flows southward down the coasts of Labrador and Newfoundland. The icy water stays close to the American coastline. Some of it comes as far south as Cape Cod. From here south it gradually disappears beneath the warmer Gulf Stream. One of the heaviest fog banks in the world lies over the cold water of the Labrador Current. The fog is the result of the cooling of the moist air by the waters of the current.

Remember when you swim in the Atlantic Ocean that the coldness of the water is caused by the Labrador Current that is between you and the Gulf Stream farther out in the ocean.

The Atlantic currents of the Southern Hemisphere are very similar to those of the north. The currents move counterclockwise, to the west, south, east, north. The heaviest current is in the eastern, instead of the western, part of the ocean. This cold current, called the Benguela, moves along the west coast of Africa. The South Equatorial Current loses some of its water to the North Atlantic. The remainder of the current becomes the Brazil Current. This circles south and then east as the South Atlantic or Antarctic Current.

The Pacific Ocean, the vast stretch of water reaching from the South Pole up the western side of South and North America is the largest ocean in the world. The word *pacific* means calm or gentle, but this ocean, which covers one third of the surface of the earth, is often rough, churning water.

The area the Pacific Ocean covers is so large that sections of it are called by compass points or by place names to show the various parts. Often you will find references to the north Pacific Ocean, the central Pacific Ocean, or the western Pacific Ocean. Sometimes the South Pacific is referred to as if it were a place all its own, while it really is part of the whole Pacific Ocean.

The currents that flow in the Pacific Ocean are quite similar in many ways to the Atlantic Ocean currents. Some of them even have the same names as those of the Atlantic.

North of the equator we find the North Equatorial Current, which originates off the coast of Mexico near the Revilla Gigedo Islands. It is the longest westerly running current on earth. A broad and fairly deep current, it flows across the entire width of the Pacific Ocean to the Philippine Islands, a distance of about 7,500 miles. It travels about twelve to eighteen miles a day. Branches from this current go north and south and become other currents.

The larger part of the North Equatorial Current, after it has passed the Mariana Islands, flows northward toward the eastern coast of Taiwan (Formosa). Then it heads for the southwestern coast of Japan by a route west of Nansei Shoto. It forms a current which is sometimes called the Japan Stream or Current. It is also called by the Japanese name of *Kuroshio*. This means "black stream." The name comes from the water of the current, which is such a deep, indigo blue that it looks black.

Actually a continuation of the North Equatorial Current, the Kuroshio is known as the Gulf Stream of the Pacific because in many ways it is quite like the Atlantic Gulf Stream. The Kuroshio is a warm current and the temperature of the water is a guide to the location of the current.

To the east of Yokohama, Japan, the current separates. One part turns north and becomes the Tsushima Current. The other part of the stream, like the Gulf Stream of the Atlantic, turns eastward and flows across the Pacific Ocean. It continues as an easterly flowing drift current. Turning southward off the coast of Vancouver Island in Canada, it becomes the California Current, flowing along the California coast as a cold current, and finally completing the Pacific circle by joining the North Equatorial Current near the Gulf of California.

South of the equator and flowing up the western coast of South America is the Humboldt Current. Flowing north, it joins the South Equatorial Current near the coast of Peru. This river of the sea travels about twenty-four miles a day. Thor Heyerdahl with five friends rode this current for 101 days on his raft *Kon Tiki* in 1947. They drifted 4,300 miles driven by the trade winds to prove a theory that originally the people of the South Pacific islands could have come from South America. His book about the voyage of *Kon Tiki* is an exciting adventure story.

More recently, in 1957 a seventy-year-old man, William Willis, spent 130 days on his raft. He called his raft *Age Unlimited*. He sailed with the Humboldt Current as it swept up the coast from the Antarctic Ocean until it joined the South Equatorial Current and then continued westward. Willis had hoped to reach Australia, but his craft became damaged and he was driven ashore at Samoa.

The Humboldt Current was named for Baron Alexander von Humboldt, a German scientist who discovered the current more than 160 years ago. He made an early record of its temperature in 1802.

Humboldt was very interested in everything about seacoasts, whether it was birds, trees, or sea water. While Humboldt was in the Andes Mountains in South America, he studied the effects that altitude, or the height of the mountains, had on climate. Later Humboldt journeyed down the mountains to the Pacific Ocean. He was amazed at the little rainfall that occurred along this part of the Pacific coast. His interest in climate and in seacoasts resulted in his discovery of the cool current that bears his name.

The Humboldt Current is sometimes called the Peru Current because of its nearness and great value to the country of Peru. The Humboldt Current is completed to the south by what is called the West Wind Drift. The old sailing captains called these windy regions the "roaring forties."

Millions of tiny plants live in the cold Humboldt or Peru Current. These supply food for small animals, which are in turn food for larger animals. Sea birds feast upon the fish and return to the islands off the coast of Peru (near the current) for nesting and roosting. Thousands of tons of high-grade guano are developed from the droppings of the birds. This is used for fertilizer.

Every so often (fortunately it does not usually happen more than once in ten years) the life-giving flow of the

Humboldt Current moves out to sea. Then the temperature of the water near the shore of Peru becomes very warm. This occurrence is called *El Niño* because it happens near Christmas, the birthday of El Niño, the Christ Child.

During the time of El Niño the fish die from the effect of the warm water. Fishermen have a difficult time catching enough fish to earn a living. The guano birds do not have fish to eat. They leave their nesting places to find food, or die of starvation on the nearby beaches.

While the shift in the Humboldt Current is perhaps the more well known, the same type of phenomenon is known in other parts of the world as well. Coastal regions of California, southwest Africa, western Australia, and Vietnam have reported similar changes in currents. The results have not been as drastic as those caused by the shift in the Humboldt Current.

Shifts in currents and increase in temperature of the water are sometimes accompanied by an invasion of "red tides." These are a beach problem in California, in Florida, and along the Gulf Coast. The change in the water is caused by the sudden appearance of tiny organisms. During the summer they multiply rapidly. These organisms send out a slimy substance which robs the water of oxygen, killing fish. Swimmers suffer no ill effect from the slightly sticky stuff, but in some areas swimming is prohibited during the red-tide season. The part that is most disagreeable is the smell of the dead fish along the beach and in the water. During the season of the red tide (the time they last varies), fishing and gathering of mussels and other shellfish is prohibited. The seasonal increase of these organisms has been known since ancient times. It is thought the Red Sea owes its name to a red tide in an enclosed sea.

In comparison with the knowledge we have of the Atlantic and Pacific Oceans, less is known about the Indian Ocean. However, today scientists are spending long periods of time studying the Indian Ocean. In future years our knowledge will be greatly increased by the findings of these men and women.

While the Indian Ocean is smaller in area than the Pacific Ocean, the currents of the Indian Ocean are equally as interesting. They are complicated by the *monsoons*. Monsoons are winds that develop over the northern Indian Ocean and the lands that are to the north and to the east.

Many people think of monsoons only as the rainy season that accompanies the winds. However, a monsoon is a wind. Look the word up in the dictionary. You will find the first use given is that of "a wind."

During the winter the winds move from the cold mountain area out over the Indian Ocean. They strengthen the prevailing northeast trades. This is called the winter monsoon.

During the summer the wind blows from the Indian Ocean in a reverse direction and becomes a southwest wind. This is the summer monsoon. The directions of the winds

Summer or southwest monsoon   *Winter or northeast monsoon*

would usually be north and south. They are deflected north-east and southwest by the spinning of the earth.

These winds cause changes in currents as well. The currents of the Indian Ocean are reversed with the change of the monsoon winds.

While we know about currents in the Atlantic, Pacific, and Indian Oceans, there is much yet to be learned.

Several years ago, during the months from July 1957 to December 1958, scientists and governments of fifty-four nations agreed to cooperate in a world-wide program of scientific investigation. This period of study was called the International Geophysical Year. Very often it is shortened and just called IGY. During these months scientists made discoveries about the earth, the floor of the sea, and the flowing of the currents, as well as adding greatly to other scientific findings. It will take years before the information gathered by these fifty-four nations can be analyzed and be made available. The IGY was a splendid research project, for each group agreed to share its findings with all of the others. Even representatives of nations who were not very friendly joined in this cooperative effort to increase our knowledge of the earth.

During the IGY a British-American team located a fairly large current nearly *underneath* the Gulf Stream. It was found at a depth of 6,600 to 9,000 feet and was flowing in the opposite direction to the Gulf Stream. As yet scientists have not mapped it in detail.

Look at a world map in an atlas. See if you can find longitude 150 degrees west on the equator. You will probably find the words "International Dateline" at 180 degrees longitude, and on most maps you will see the Marshall Islands, the Gilbert Islands, and the Phoenix Islands nearby to the west. It was near this location in 1952 that Townsend Cromwell was working for the United States Fish and Wildlife Service. He was experimenting with Japanese long-line fishing methods. He had let out several lines of cable with smaller fishing lines attached. Since Cromwell was in the drift of the South Equatorial Current, he thought that his line would move toward the west or away from the United States. Instead it headed rapidly eastward toward his home in San Diego. This was the first time that anyone had even thought that there was a large underwater current running in the *opposite* direction from the surface current.

Shortly after this, Cromwell was killed in an air crash, but the current he had found was named for him. During the IGY, John Knauss of the Scripps Institution of Ocean-ography headed a group that traced the Cromwell Current halfway across the Pacific Ocean. They found it to be as big and powerful as the Gulf Stream.

Today we know that the current runs for 3,500 miles. At the present time it is thought that the current disappears at the Galapagos Islands off the coast of Ecuador. The current itself is about 250 miles wide. It has a speed of about 3.5 miles an hour. It is found several hundred feet under the South Equatorial Current.

On this same trip Knauss and his expedition found a third current. This was a much weaker current that flowed westward again at still greater depths.

Even though the experts have much to learn about surface currents, there is still more to learn of the underwater or subsurface currents. Some scientists think that they are one of the most exciting discoveries of this generation and that the study of these currents will make a great deal of difference in our understanding of weather and climate.

When the atomic submarine *Thresher* sank in 1963 with the loss of 129 men, the first investigations showed that there was a possibility that an unknown deep-water current could have had some influence on the disaster. Although even today we do not know all the answers, perhaps someday we will know what actually happened as oceanographers learn more about currents under other currents, and of deep underwater waves.

A world's championship surfer competing in Australia in 1964.

CHAPTER VII

Waves, Tides, and Currents of Tomorrow

Along the warm shores of both the Atlantic and Pacific coasts, people gather during the summer months. In the milder climates of the southern parts of the coasts, they vacation all winter long. Many people make their permanent homes there as well.

To some of these people the ever-changing sea is interesting to watch. To many it is a wonderful place to play, and to others the seashore is an excellent place to live as well as a vantage point from which to study the sea itself.

People who live by the sea have come to realize the dangers that sudden changes in weather can bring. They have learned as well that waves, tides, and currents can all be dangerous if they are not understood.

Surfboard riders, balancing their boards on the advancing waves, must acquire split-second timing to keep their boards at the breaking point of the wave. Then they can ride close in to shore. They have learned never to surf-board alone, and they know the need to be good strong swimmers who understand waves, tides, and currents.

Surfboarders soon learn to watch for *longshore currents*, and the *rip currents* that so often take swimmers by surprise. Water that is brought to shore by waves must go somewhere. It tends to run along the shore, and so creates a longshore current, until it heads seaward at a place where incoming waves are weaker. Sometimes this return to deeper water creates a fast-moving current known as a *rip*. Often it is misnamed a *rip tide* or an undertow.

Scientists believe that there are no undertow currents that physically pull a swimmer down. Near shore the rip currents extend to the bottom. Farther out to sea they become surface currents. These rip currents can prove dangerous to swimmers.

If you should be caught in a rip current, do not fight it and do not become frightened. A rip current is never very wide. Drift with the current, or swim across it until you reach slack water in a more shallow part.

One of the best ways to avoid a rip current is to know the coast where you are swimming. Obey the signs that are posted to warn you of rip currents, and do not swim in these areas.

If you are a strong swimmer and a surfboarder and understand rip currents, the rips can give you a free ride

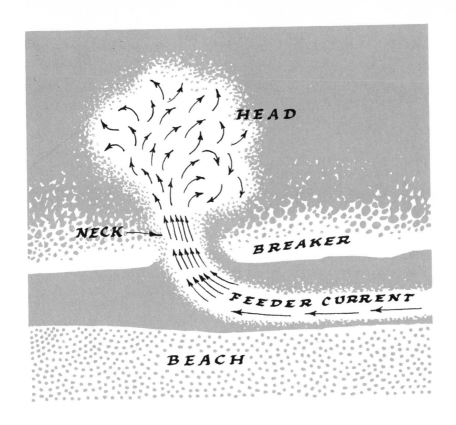

HEAD

NECK ⟶

BREAKER

FEEDER CURRENT

BEACH

out through the breakers. This, however, is not for a beginner to try. Even experienced swimmers and owners of small boats are often caught off guard by a sudden shift in a rip current.

Children and adults are sometimes caught by these changes even while wading in the surf. Be cautious about any strange seacoast.

Innocent-looking rip currents such as the one at Baja California, a part of Mexico, can be dangerous to swimmers.

Waves, particularly storm waves, break and grind material and carry it with them to shore. Along the Atlantic coast in summer the average blow of a breaker is about 600 pounds on every square foot of surface. In winter this can rise to almost 6,000 pounds per square foot. Compare these figures with Thomas Stevenson's report of storm waves in Scotland (p. 64).

If a wave has collected sand, pebbles, or larger rocks and has this weight behind it, you can certainly see why the wearing or washing away, called *erosion*, along shorelines is so great during heavy storms.

The March 1962 storm along the Atlantic coast and other similar disturbances have changed the shape of beaches overnight. Beaches and land along cliffs are in constant danger from erosion.

Aside from storms, the blocking or changing of the flow of currents can cause great damage. In Monterey Bay in central California, a breakwater was built to protect the small fishing craft that sail out of the harbor each morning for deep-sea fishing. Later another breakwater was built in a cross direction to provide protection and a docking area

During a storm in 1962, telephone poles on the beach at Assateague Beach, Virginia, were uprooted and washed away.

for small pleasure boats. However, by changing the flow of water toward shore and blocking other areas, the breakwaters caused large amounts of sand to pile up in the small pleasure-boat harbor, and huge amounts of beach and parking areas for cars were washed from the shore. Only by very careful planning and the spending of large amounts of money was the harbor for the pleasure boats made usable again.

Many other harbors along the Pacific coast have had difficulty, too. Because so many people are using small boats, towns have built safe harbors for them. In Santa Barbara, the town built a breakwater. The currents swirled in around this and dropped the sand they were carrying. Men near the harbor found that about 800 cubic yards of sand came into the harbor in one day! Now they have had to maintain a permanent dredge. This pumps sand from the bottom through long pipes out beyond the harbor. This keeps the harbor free of sand until the currents bring it in again.

In Florida, overcoming erosion is a problem as well. Near Pompano Beach the coastline has been washed away and houses have been undermined and threatened with destruction. To overcome this hazard, man-made bulkheads or *groins* (wooden or rock structures that are built into the ocean to protect the land) allow the sand to build up against the groin boards or rocks.

Scientists and engineers have had different opinions about groins. Some have favored long ones. Others have

Groins built in some areas help prevent erosion of beach.

thought that short groins were more effective. Some recommend solid structures made of various materials, while others thought it best to have them open. In this way water could flow through the groins, and supposedly some sand would stay behind. Some scientists believe that groins in many cases make erosion worse.

Today engineers consider all factors before making any recommendations to cure erosion. They often actually build beaches. This is sometimes called "beach nourishment." This technique, being used along the Florida coast, is similar to the dredging being done in Santa Barbara Harbor but it is done to build up the beaches rather than to make a harbor deeper.

Tons of sand are dredged from bays and harbors and pumped along the shores of the ocean. Often sand is brought from inland by truck and dumped on beaches. If the work is done properly, much of the sand stays and builds up the beaches.

Men are learning that a straight up-and-down wall can be more easily undermined by the waves, while a sloping wall, similar to a sloping beach, can take more punishment from the pounding waves.

Erosion of a cliff can be stopped temporarily at least, although it is an expensive project. By bringing stones, big, broken, rough pieces of concrete, and other heavy material and placing them along the lower part of the cliff, the pounding from the waves against the cliff is lessened.

Naturally, when waves are heightened by wind their force is increased. Approximate estimates of the strength of the wind may be shown by the effect of wind on objects. These estimates were developed by Admiral Sir Francis Beaufort (pronounced Bo-fort), who first introduced them into the English Navy in 1805.

Admiral Beaufort used seventeen numbers in his scale, although the numbers above twelve show only the intensity of hurricane-force wind. The following table shows the Beaufort number, indicates the words used to name the wind, gives the miles per hour, and tells the distinctive char-

Waves can be duplicated and studied in wave channels. The device can be adjusted to nearly any wave and shore situation.

acteristics of the wind. In addition to the usual description of the wind, this table gives information not usually found on the Beaufort Scale. It describes what you might see on the sea. For example, look at the first listing. Under the description of the Beaufort Number 0, you will find: "Calm, smoke rises vertically." Then the additional information is given: "Sea like a mirror." This will help you understand the Beaufort Scale in connection with the sea.

BEAUFORT NUMBER	NAME OF WIND	MILES PER HOUR	DESCRIPTION OF WIND AND SEA
0	Calm	Less than 1	Calm; smoke rises vertically; sea like a mirror.
1	Light air	1 to 3	Direction of wind shown by smoke but not by wind vanes; ripples with appearance of scales are formed. without foam crests.
2	Light breeze	4 to 7	Wind felt on face; leaves rustle; ordinary vane moved by wind; small wavelets still short but more pronounced; crests have a glassy appearance and do not break.
3	Gentle breeze	8 to 12	Leaves and small twigs in constant motion; wind extends light flag; large wavelets, crests begin to break; foam of glassy appearance; scattered whitecaps.
4	Moderate breeze	13 to 18	Raises dust and loose paper, small branches are moved; small waves becoming longer; fairly frequent whitecaps.
5	Fresh breeze	19 to 24	Small trees in leaf begin to sway; crested wavelets form on inland water; moderate waves, taking on a more pronounced long form on open sea; many whitecaps; chance of some spray.
6	Strong breeze	25 to 31	Large branches in motion; telegraph wires whistle; umbrella used with difficulty; large waves begin to form; extensive white foam crests; probably some spray.
7	Moderate gale	32 to 38	Whole trees in motion; inconvenience in walking against wind; sea heaps up; white foam from breaking waves begins to blow in streaks along direction of wind; spray begins to be seen.

BEAUFORT NUMBER	NAME OF WIND	MILES PER HOUR	DESCRIPTION OF WIND AND SEA
8	Fresh gale	39 to 46	Breaks twigs off trees; generally impedes progress; moderately high waves of greater length; edges of crest break into spray; foam blows in well-marked streaks along direction of wind.
9	Strong gale	47 to 54	Slight structural damage occurs; chimney pots and slates removed; high waves; dense streaks of foam along direction of wind; sea begins to roll; spray affects visibility.
10	Whole gale	55 to 63	Trees uprooted; considerable structural damage occurs; very high waves with long overhanging crests; foam makes dense white streaks along direction of wind; sea takes on white appearance; rolling of sea becomes heavy; visibility affected.
11	Storm	64 to 72	Very rarely experienced; accompanied by widespread damage; exceptionally high waves; small and medium-sized ships may be lost to view behind waves; sea covered with long white patches of foam; edges of wave crests blown into foam; visibility affected.
12	Hurricane	73 to 82	Devastation occurs; air filled with foam and spray; sea is filled with driving spray; visibility very seriously affected.
13	Hurricane	83 to 92	Devastation occurs.
14	Hurricane	93 to 103	Devastation occurs.
15	Hurricane	104 to 114	Devastation occurs.
16	Hurricane	115 to 125	Devastation occurs.
17	Hurricane	126 to 136	Devastation occurs.

When next you see large branches in motion, or hear whistling in telegraph wires, you will know that somewhere on some weather chart some person is marking the number 6, and adding "strong breeze, 25 to 31 miles per hour." Perhaps you will hear an announcer giving this information over the radio or television. Wind information is essential to many persons throughout the country. Many families listen to several weather broadcasts a day. Think of the damage that wind can do inland as well as along our sea-shores.

An ambitious project that may be developed during your lifetime is the harnessing of the energy of the waves to provide electric power.

In the daily tides of the Bay of Fundy, scientists have said that an estimated 200 million horsepower could be generated if the tides were used. Schemes for a tidal power plant on Passamaquoddy Bay, between Maine and New Brunswick at the mouth of the Bay of Fundy, have been discussed for years. It was first suggested by Dexter P. Cooper in 1919.

Ten years ago a group of men from the United States and Canada studied the possibility of harnessing the power of the tides. Their plan was to trap the sea water in a natural, island-studded reservoir. The water was then to be channeled through huge penstocks. These are conduits between two gates or dams that keep the water at a certain level. When the water reached the adjacent basin (kept at low-tide level by other gates), this would power hydro-electric generators. These were to have an output of about a third that of Hoover Dam. However, little has been done with the project. The estimate of the expense has always been too great.

In France today a project is being developed that will put a tidal change of forty feet to work producing electricity. The huge project has dammed the half-mile-wide mouth

The daily tidal change has been put to work for the first time at the hydroelectric dam at Rance, France. In the top picture, the sluice gate is opened to fill the reservoirs. The dam, after it is filled, is shown in the bottom picture. Full operation of the dam was begun late in 1966.

of the Rance River in northern France. This was done so that the river bottom could be dried. Then a permanent dam, which is to contain turbines, will be constructed. The huge tidal surge will turn the turbines to drive generators during both the incoming and the outgoing tides. When the project is completed, it will be a pioneering engineering venture and will probably lead to others like it.

Perhaps in future years other engineering feats will be accomplished to make waves, tides, and currents work for us. Stranger things have occurred in our lifetime, and perhaps you will have a part in making them happen.

With the astronauts not only flying and walking in space, but also working with a team on underwater research, many new and exciting discoveries are taking place.

In September 1965, half a mile off the coast of La Jolla in southern California, Commander Scott Carpenter, one of the original astronauts, spent a month 205 feet below the surface of the ocean. He became the first man to become an aquanaut as well as an astronaut.

Using Scripps Institution of Oceanography as a base and living in *Sealab II*, the United States Navy's capsule in inner space, with a crew of ten men, Carpenter spent an entire month below the surface of the ocean, although the crew changed each two weeks.

The aquanauts' time in a steel cylinder that measured twelve feet by fifty feet proved that men can live and work together below the surface of the ocean. The men set up an outdoor station to measure ocean currents and performed about one hundred experiments in marine biology and oceanography.

Their first steps from *Sealab II* into the inky black so far below the surface were terrifying. At 50 degrees Fahrenheit the water felt extremely cold despite their special rubber suits. The men found they could get lost twenty feet from the capsule!

One of the interesting items in the month's experiments was the part a trained porpoise named Tuffy played in the work the men were doing. Tuffy lived in a pen near the support barge on the surface and carried messages and cables between *Sealab II*, the divers, and the surface or "topside."

Tuffy was a seven-foot, 270-pound bottlenose porpoise. He had been taught to respond to a buzzer. When the buzzer was sounded, he came swimming toward it. Often it brought him to the men working outside *Sealab II*. Then Tuffy would lead the men back to home base. Although the men became confused as to direction, Tuffy could always find his way home!

One month later, in October 1965, off Cap Ferrat on the French Riviera, a group of six Frenchmen lived in a yellow-and-black checkerboard underwater "house." They were anchored at a depth of 330 feet and worked even deeper, going to a depth of 375 feet. The oceanauts' work was commanded by Captain Jacques Cousteau, the pioneering French underwater explorer. He directed the activities of the three-week mission of *Con Shelf III* (named for the Continental Shelf) from a lighthouse on shore. The experiments made during the mission were developed specifically to plan methods to use in working with undersea deposits of oil.

During this same month off the coast of South Carolina, scientists from the Woods Hole Oceanographic Institution installed the first stable deep-sea structure in water a half-mile deep. Named *Sea Spider* and designed by Godfrey H. Savage of Woods Hole, it is a saucer-shaped aluminum float, securely held to the ocean's bottom by four long steel cables. Various instruments and buoyant glass spheres are attached along the spider-leg cables. The saucer is placed 110 feet below the ocean's surface to avoid buffeting by wind and waves.

A telemetering buoy at the sea surface transmits data, collected from the instruments by radio, to a nearby oceanographic vessel. The instruments record such measurements as the speed of ocean currents, temperature variations of the water, and underwater sounds.

You may not believe it, but the sea is a noisy place! Scientists have found that the sounds made by animals in the sea compare roughly in intensity to the hum of a busy office where typewriters are clattering, telephones ringing, and people talking.

A recent study of the not-so-silent sea was undertaken by the American Museum of Natural History for the United States Naval Training Device Center in Port Washington, New York. It was a study made to enable Navy sonar operators to tell which noises were made by sea life and which came from unknown ships, submarines, or other craft.

The noises made by numerous marine animals are often in frequency ranges that can interfere with sonar and underwater listening equipment. Surface waves or deep currents provide other sounds of the sea. Part of the problem of any such study is the fact that sound travels almost five times faster in the sea than it does in the air!

Industry is working with science to find out more and more about oceans. Earlier in 1965 off the Santa Barbara coast near Goleta Harbor, a 136-foot converted minesweeper named *Swan* went almost 900 miles out to sea from her home port. She met a tug called *Horizon* from the Scripps Institution of Oceanography at La Jolla. There in the sea, miles from shore, scientists and their assistants worked with a number of scientific devices to track patterns made by sound through something called the "surface duct." This is a layer of water between 100 and 1,000 feet down.

For almost two weeks, tracking back and forth, the scientists tallied wave heights, sea temperatures, and water pressures at different depths. All of their findings were correlated with surface weather data in an effort to find if there was any connection between two or more of these items.

Another "first," this information was tabulated by a computer on board one of the ships. An aircraft corporation and one of the major automotive industries provided the research devices on board the ships.

Today the whole world is becoming more and more interested in the sea that surrounds us, and in the information and help that can come from waves, tides, and currents. Men and women in science are keenly aware of what oceanography can mean to us. Business men are financing investigations of the oceans, for they realize how much the waters of the seas can give to the people of the earth.

Many persons feel that money for discoveries in space could be better spent in developing the depths of the waters and harnessing the currents; others suggest that space and ocean explorations should go hand in hand.

Oceanography development is one of the great future needs of the world. You may find yourself involved in sea discoveries. Who knows what part you may play, who knows what you may find, or how you will work with the resources, or how you will solve the mysteries of the seas. Perhaps your work, whether in business or in science, may be one of the great contributions to the world in which we live with waves, tides, and currents.

GLOSSARY

Axis—an imaginary line that runs through the center of the earth about which the earth turns.

Basins—pool-like places on the ocean floor that vary in depth and size; usually circular, elliptical, or oval in shape.

Bore—a wall of rushing water.

Breaker—an individual breaking wave.

Coriolis effect—the tendency for motion to be deflected to the right in the Northern Hemisphere; to the left in the Southern Hemisphere.

Crest—the top or highest part of a wave.

Currents—the rivers of the seas.

Diurnal tide—having only one high water and one low water each tidal day.

Ebb tide—tide water that moves away from shore.

Fetch—the distance wind has blown without changing direction.

Flood tide—incoming rising water.

Gravity—an invisible force that holds the moon, the earth, the sun, and other objects in position.

Ground swell—a swell when it reaches quite shallow water. See *swell*.

Gyre—a circular motion described by a moving object.

High tide—the highest point of rising tide.

Longshore current—a current located in the surf zone; water that runs along shore. See *rip current, undertow.*

Mixed tide—having differences in either high- or low-water heights, with two high waters and two low waters usually occurring each tidal day.

Monsoon—a wind, especially in the Indian Ocean and southeastern Asia; also the rainy season that accompanies the wind.

Neap tide—tide with smallest range. See *range*.

Orbit—the course followed by one heavenly body around another, as the earth around the sun; to travel in circles, as forward motion of water; used commonly today because of space travel.

Oscillation—a back-and-forth or up-and-down motion.

Range—the difference between the water levels at high tide and low tide.

Rip current—a strong surface current of short duration flowing away from the shore. See *longshore current, undertow.*

Seismic sea wave—same as *tsunami.*

Semidiurnal tide—having two high waters and two low waters each tidal day.

Storm tide—rise of water level during time of storm waves.

Storm waves—caused by high winds that blow toward shore.

Surf—the region of breaking waves, including foam and spray; collective term for breakers.

Swell—a wave activity where there is no wind.

Tidal wave—incorrect name for *seismic sea wave* or *tsunami.* See *tsunami.*

Trade wind—steady wind that blows from east to west.

Trough—the lowest part of the wave between crests.

Tsunami—(pronounced tsū-nah′-mē) a progressive gravity wave caused by earthquakes or other underwater disturbances; commonly misnamed "tidal wave"; often called *seismic sea wave.*

Undertow—an incorrect name for longshore or rip current. See *longshore current; rip current.*

Universal law of gravity—the force of attraction of two objects for each other; the force depends upon the mass (weight) of each body and the distance between the bodies.

Wind wave—a wave produced by action of the wind; a ripple started by a breeze or wind that develops into a full-size wave under certain conditions.

BIBLIOGRAPHY

Bascom, Willard: *Waves and Beaches*. New York: Doubleday and Company; 1964.

Beauchamp, Wilbur L., and Helen J. Challand: *Basic Science Handbook*. Glenview, Illinois: Scott, Foresman and Company; 1961.

Beckinsale, Robert P.: *Land, Air, and Ocean*. London: Gerald Duckworth and Company, Ltd.; 1960.

Carrington, Richard: *A Biography of the Sea*. New York: Basic Books, Inc.; 1960.

Carson, Rachel L.: *The Edge of the Sea*. Boston: Houghton Mifflin Company; 1960.
The Sea Around Us. New York: Oxford University Press; 1950.

Cowen, Robert C.: *Frontiers of the Sea*. New York: Doubleday and Company; 1960.

Defant, Albert: *Ebb and Flow*. Ann Arbor: University of Michigan Press; 1958.

Del Rey, Lester: *The Mysterious Sea*. Philadelphia: Chilton Company; 1961.

Freuchen, Peter: *Book of the Seven Seas*. New York: Julian Messner, Inc.; 1958.

Hedgpeth, Joel W.: *Seashore Life*. Berkeley: University of California Press; 1962.

Irving, Robert: *Hurricanes and Twisters*. New York: Alfred A. Knopf; 1955.
Volcanoes and Earthquakes. New York: Alfred A. Knopf; 1962.

Jenkins, J. T.: *A Text Book of Oceanography*. New York: E. P. Dutton and Company; 1935.

Leip, Hans: *River in the Sea, The Story of the Gulf Stream.* New York: G. P. Putnam's Sons; 1958.

Outhwaite, Leonard: *The Atlantic, A History of an Ocean.* New York: Coward-McCann, Inc.; 1957.

Perry, John, and Jane Greverus: *Exploring the Seacoast.* New York: Whittlesey House, McGraw-Hill Book Co., Inc.; 1961.

Ricketts, Edward F., and Jack Calvin: *Between Pacific Tides.* Stanford, California: Stanford University Press; 1952.

Shepard, Francis P.: *The Earth Beneath the Sea.* Baltimore: The Johns Hopkins Press; 1959.

Stommel, Henry: *The Gulf Stream.* Berkeley: University of California Press; 1958.

Sverdrup, H. U., Martin W. Johnson, and Richard H. Fleming: *The Oceans.* Englewood Cliffs, New Jersey: Prentice-Hall; 1942.

Grateful acknowledgment is made for permission to use the following photographs:

Alpha Photo Associates, 42

Black Star Publishing Company, Flip Schulke, 61 bottom; Vernon D. Sutcher, 5 top; Fred Ward, 91; Ace Williams, 39; Werner Wolfe, 72

Lee Blaisdell, 2

Wynn Bullock, front endpapers, 5 bottom, 18

Edward C. Forsyth, 7

Ewing Galloway, 103, back endpapers

Long Beach Island Board of Trade, 61 top

Monkmeyer Press Photo Service, 35 top, 35 bottom; Pro Pix, 26

U.S. Coast Guard Official Photographs, 59, 89

U.S. Department of Commerce, Environmental Science Services Administration, Coast and Geodetic Survey, 25, 55 top, 88

U.S. Navy Official Photograph by C. E. Wall, 53

Wide World Photos, Inc., 10, 51, 55 bottom, 84, 93, 97

INDEX

Antarctic Current, 75
Arctic Current, 74-75
Atlantic Currents, 70-75
axis (earth's), 9-14, 20, 22, 68-69

Bay of Fundy, 33-36, 96; *see also* bore
Beaufort Scale, 93-95
Benguela Current, 75
bore, 34-37
Brazil Current, 75
breaker, 50; *see also* waves
breakwater, 89-90

California Current, 77
Canaries Current, 74
Carpenter, Scott, 98
Con Shelf III, 100
Coriolis effect, 68-69, 74
Cousteau, Jacques, 99-100
crest, 46; *see also* waves
Cromwell Current, 82-83
Currents, 65-83; *see also* rip, longshore currents. Antarctic Current, Arctic Current, Atlantic Currents, Benguela Current, Brazil Current, California Current, Canaries Current, Cromwell Current, eastern drift, Florida Current, Gulf Stream, Humboldt Current, Japan Current, Labrador Current, North Atlantic Drift, North Equatorial Current, Pacific Currents, South Atlantic Current, South Equatorial Current, Tsushima Current, West Wind Drift

dam, hydroelectric, 96-98
diurnal tide, 21

earthquakes, 52, 56-57; *see also* seismic sea waves
eastern drift, 74

ebb tide, 4
erosion (of beach), 88-92

Ferrel's law, 68; *see also* Coriolis effect
fetch, 47; *see also* waves
flood tide, 4, 40
Florida Current, *see* Gulf Stream
Franklin, Benjamin, 71-73

gravity, 14-22
groins, 90-92
ground swell, *see* swell
Gulf Stream, 70-74, 82
gyre, 65; *see also* currents

half tide, 35
Heyerdahl, Thor, 77
high tide, 3-8; *see also* range of the tides
Humboldt Current, 77-79

International Geophysical Year (IGY), 81-83

Japan Current, 76-77

Kuroshio, *see* Japan Current

Labrador Current, 74-75
longshore currents, 86
low tide, 3-8; *see also* range of the tides

minimum range, *see* neap tides
mixed tides, 28
monsoons, 80-81
moon, 14, 17, 19-22

neap tides, 22
Newton, Sir Isaac, 15-16

"noises" of the sea, 100-101
North Atlantic Drift, 74
North Equatorial Current, 70, 74 (Atlantic); 76-77 (Pacific)

Oceans: basins, 27; Antarctic, 66, 69, 73; Arctic, 66-69; Atlantic, 66,
   70, 73-75; Indian, 66, 80-81; Pacific, 66, 75-80
orbit (earth's), 9-14, 21

Pacific Currents, 75-80
Peru Current, *see* Humboldt Current

range of the tides, 4-8, 27-43; *see also* bore
"red tides," 79-80
rip, 86
rip currents, 86-87
"rivers of the sea," *see* currents
rotation (earth's), 9, 12-14; *see also* axis

Scripps Institution of Oceanography, 83, 98, 101
Sealab II, 98-99
seasons (causes of), 10-12
seismic sea wave, 52-58
semidiurnal tide, 20-28
South Atlantic Current, 75
South Equatorial Current, 70, 75 (Atlantic); 77 (Pacific)
spring tides, 22, 36
storm damage, 60-64
storm waves, 60, 88
sun, 9-12, 16-17, 20-22
surf, 50; *see also* waves
swell, 50; *see also* waves

tidal bulge, 20-21
tidal wave, 36
tide gauge, 31-33
tide-predicting machines, 22-24

tide tables, 29-31, 36, 38
tides, *see* bore, diurnal tide, ebb tide, flood tide, half tide, high tide, low tide, mixed tides, range of the tides, semidiurnal tide, spring tides, tidal bulge, tidal wave, tide gauge, tide tables
trade winds, 68-70
trough, 46; *see also* waves
tsunami, *see* seismic sea wave
Tsushima Current, 77

undertow, *see* rip
United States Coast and Geodetic Survey, 23, 57

wave dynamometer, 64
wave motion, 44-52
waves, 43-64; formation, 43-45; *see also* breaker, crest, fetch, seismic sea wave, storm damage, storm waves, surf, swell, trough, wave dynamometer, wave motion, wind wave
West Wind Drift, 78
wind wave, 43
Woods Hole Oceanographic Institute, 58, 100

A BOUT THE A UTHOR

"THERE ARE FEW TIMES in my life when I have lived far from the sea," writes Elizabeth Clemons. A third generation Californian, she lives in Carmel Valley, only fourteen miles from the ocean.

During the four years she was writing *Waves, Tides, and Currents,* she and her husband traveled oceans and seas—even including a trip to Korea to study the turn of the tide. Many of the examples in the book have come from personal experiences with waves, tides, and currents, in rivers, harbors, and open seas.

Elizabeth Clemons has been a teacher of teachers, of college students, and of elementary school children. She has also been an editor of juvenile books. This is her third book for Knopf.

A Note on the Type

THIS BOOK WAS SET *on the Lintoype in Janson, a recutting made direct from the type cast from matrices made by Anton Janson. Whether or not Janson was of Dutch ancestry is not known, but it is known that he purchased a foundry and was a practicing type-founder in Leipzig during the years 1660 to 1687. Janson's first specimen sheet was issued in 1675. His successor issued a specimen sheet showing all of the Janson types in 1689.*

His type is an excellent example of the influential and sturdy Dutch types that prevailed in England prior to the development by William Caslon of his own incomparable designs, which he evolved from these Dutch faces. The Dutch in their turn had been influenced by Garamond in France. The general tone of Janson, however, is darker than Garamond and has a sturdiness and substance quite different from its predecessors. It is a highly legible type, and its individual letters have a pleasing variety of design. Its heavy and light strokes make it sharp and clear, and the full-page effect is characterful and harmonious.